Amphibionics

Amphibionics

Build Your Own Biologically Inspired Robot

Karl Williams

McGraw-Hill

New York   Chicago   San Francisco   Lisbon
London   Madrid   Mexico City   Milan
New Delhi   San Juan   Seoul
Singapore   Sydney   Toronto

*The **McGraw-Hill** Companies*

Cataloging-in-Publication Data is on file with the Library of Congress

Copyright © 2003 by The McGraw-Hill Companies, Inc. All rights reserved. Printed in the United States of America. Except as permitted under the United States Copyright Act of 1976, no part of this publication may be reproduced or distributed in any form or by any means, or stored in a data base or retrieval system, without the prior written permission of the publisher.

1 2 3 4 5 6 7 8 9 0 DOC/DOC 0 9 8 7 6 5 4 3

ISBN 0-07-141245-X

The sponsoring editor for this book was Judy Bass and the production supervisor was Sherri Souffrance. It was set in Tiepolo Book by Patricia Wallenburg.

Printed and bound by RR Donnelly.

McGraw-Hill books are available at special quantity discounts to use as premiums and sales promotions, or for use in corporate training programs. For more information, please write to the Director of Special Sales, McGraw-Hill Professional, Two Penn Plaza, New York, NY 10121-2298. Or contact your local bookstore.

 This book is printed on recycled, acid-free paper containing a minimum of 50 percent recycled, de-inked fiber.

To Laurie

Summary of Contents

Introduction   xv

Acknowledgments   xvii

1   **Tools, Test Equipment, and Materials**   1

2   **Printed Circuit Board Fabrication**   17

3   **Microcontrollers and PIC Programming**   25

4   **Frogbotic: Build Your Own Robotic Frog**   51

5   **Serpentronic: Build Your Own
Robotic Snake**   117

6   **Crocobot: Build Your Own
Robotic Crocodile**   191

7   **Turtletron: Build Your Own
Robotic Turtle**   271

8   Taking It Further   345

Bibliography   349

Index   351

Contents

Introduction   xv

Acknowledgments   xvii

1   Tools, Test Equipment, and Materials   1

Test Equipment   10

Construction Materials   12

Summary   15

2   Printed Circuit Board Fabrication   17

Summary   22

3   Microcontrollers and PIC Programming   25

Microcontrollers   25

PIC 16F84 MCU   26

PicBasic Pro Compiler   28

Software Installation   31

Compiling a Program   35

Using the EPIC Programmer to Program the PIC   40

Testing the Controller Board   44

MicroCode Studio Visual Integrated
Development Environment   45

Using a Programmer with MicroCode Studio   47

    MicroCode Studio in Circuit Debugger   48

Summary   49

4   Frogbotic: Build Your Own Robotic Frog   51

Frogs and Toads   51

Overview of the Frogbotic Project   52

    R/C Servo Motors   54

    Modifying Servos for Continuous Rotation   55

Controlling a Modified Servo   66

    Mechanical Construction of Frogbotic   68

Assembling the Legs   77

    Attaching the Legs to the Robot's Body   82

    Fabricating the Servo Mounts   84

    Constructing the Front Legs   90

    Leg Position Sensors   91

    Wiring the Limit Switches   91

Frogbotic's Main Controller Board   94

     Creating Frogbotic's Printed Circuit Board   96

Fabricating the Power Connector   98

Putting It All Together   100

Programming and Experiments with Frogbotic   103

5   Serpentronic: Build Your Own Robotic Snake   117

Snakes   117

Overview of the Serpentronic Project   119

     Mechanical Construction of Serpentronic   120

     Constructing the Body Sections   121

     Constructing the Tail Section   130

     Constructing the Head   132

Assembling the Snake's Mechanical Structure   137

     Connecting the Body Sections, Tail, and Head   138

Serpentronic's Main Controller Board   144

     Creating the Main Controller Printed Circuit Board   146

The Infrared Sensor Board   148

     Constructing the Infrared Sensor Circuit Board   152

Calibration   154

     Mounting the Controller and Infrared Sensor Board   155

Wiring the Robot   158

Programming and Experiments with Serpentronic   164

    Motion Control   171

Infrared Sensor   177

Summary   188

**6 Crocobot: Build Your Own
Robotic Crocodile   191**

Crocodilians   191

Overview of the Crocobot Project   193

    Mechanical Construction of Crocobot   194

    Constructing the Chassis   199

    Constructing the Body Covers and Tail Section   202

    Wiring the Limit Switches   209

    Constructing the Legs   211

    Assembling the Legs   213

The Controller Circuit Board   216

    L298 Dual Full-Bridge Driver   218

    Creating the Main Controller Printed
    Circuit Board   222

Putting It All Together   226

Constructing the Remote Control Transmitter   228

    PIC 16C71   232

Creating the Remote Control Printed
Circuit Board   234

Programming Crocobot   239

7 Turtletron: Build Your Own Robotic Turtle   271

Turtles and Tortoises   271

Overview of the Turtletron Project   272

The History of Robotic Turtles   273

Mechanical Construction of Turtletron   275

Assembling the Gearboxes and
Attaching the Wheels   277

Electronics   283

Ultrasonic Range Finding   286

The Remote Control Transmitter   298

Programming Turtletron   300

Testing the SRF04 Ultrasonic Ranger   308

Obstacle Avoidance Using the
Ultrasonic Range Finder   313

Distance Measurement Using an Optical
Shaft Encoder   325

Fabricating the Shaft Encoder   327

Room Mapping Using the Shaft Encoder
and Ultrasonic Range Finder   334

8 Taking It Further **345**

Frogbotic 345

Serpentronic 346

Crocobot 346

Turtletron 347

Bibliography 349

Index 351

Introduction

The robots in this book were designed to imitate biological life-forms. Watching the snake robot moving through a room, it is interesting to observe the surprised reactions of people when it quickly turns towards them. People actually regard the robot as being alive. I am struck with the thought that although these machines are not alive in our biological sense, they actually are alive, but as life-forms unto themselves. These artificially intelligent machines are the products of human imagination and technical understanding. As the technology advances, the line between living and non-living matter is slowly becoming blurred.

Being a collector of robotics books, old and new, I am always excited to see the robots and devices that other people have created, or interesting ways in which they have implemented various technologies and theories. I am often inspired by some of the outdated mechanical diagrams and circuits in the old robotics books. Even with today's advanced computer technology, nothing is quite as fascinating to see as the ingenious mechanical workings of a well-designed machine.

Amphibionics is a continuation on the theme of building biologically inspired robots introduced in *Insectronics*, which explored the building and experimentation of a hexapod walking insect robot. The practical research detailed in *Amphibionics* is aimed at developing a new class of biologically inspired mobile robots that exhibits much greater robustness of performance in unstructured environments than a lot of today's robots. This new class of robot is aimed at being substantially more compliant and stable than current wheeled robots.

Acknowledgments

Thanks to my parents Gordon and Ruth Williams for their encouragement. To my brothers and their wives: Doug Williams, Gylian Williams, Geoff Williams, and Margaret Sullivan-Williams. Thanks to Laurie Borowski for her love, patience, and suggestions. Thanks to Judy Bass and the team at McGraw-Hill for all of their hard work. Thanks to Patricia Wallenburg for doing a great job of putting the book together. Thanks to the following people who always have the time to discuss robotics and new ideas: James Vanderleeuw, Stacey Dineen, Sachin Rao, Chris Meidell, John Lammers, Tom Cloutier, Darryl Archer, Paul Steinbach, Jack Kesselman, Charles Cummins, Maria Cummins, Tracy Strike, Raymond Pau, Clark MacDonald, Rodi Snow, Steve Frederick Sameer Siddiqi, Dan Dubois, and Steve Rankin. Thanks to Jason Jackson, Roland Hofer, Kenn Booty, JoAnna Kleuskens, Patti Ramseyer, Myke Predko, Roger Skubowius, and Tim Jones at Cognitive Symbolics.

Amphibionics

1
Tools, Test Equipment, and Materials

During the mechanical construction phase of building the robots in this book, a number of tools will be required. You will need a workbench or sturdy table in an area with good lighting. Try to keep your work area clean and free of clutter.

The first tool that will be used is the hacksaw. The hacksaw is designed to cut metal and hard plastics. When using the hacksaw to make straight cuts, it is a good idea to use a miter box. **Figure 1.1** shows the hacksaw (labeled L) and the miter box (K).

If you have a little extra money and think that you will be building a lot of robots, then you really need a band saw fitted with a metal cutting blade. The band saw shown in **Figure 1.2** is 9 inches, meaning that the saw can cut pieces up to a maximum length of 9 inches. This is perfect for building smaller robots, like the ones detailed in this book. With the metal cutting band saw, pieces of aluminum can be cut fast and with greater accuracy than a hacksaw.

An important piece of equipment that will be needed in your workshop is a vise, like the one shown in **Figure 1.3**. The vise will be needed quite often when cutting, drilling, and bending aluminum. Always clamp metal pieces tightly in the vise when working on

FIGURE 1.1

Hacksaw and miter box.

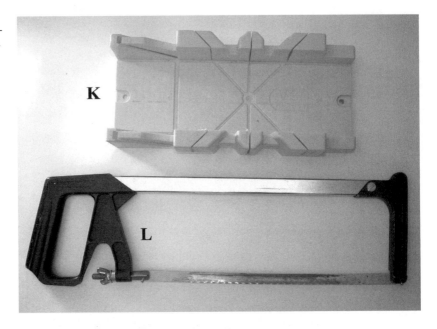

FIGURE 1.2

Band saw fitted with a metal cutting blade.

FIGURE 1.3

Work bench vise.

them with other tools. It is dangerous to try drilling metal pieces that are not clamped in a vise.

You will need an electric drill during the mechanical construction phase of building the robots and the fabrication of the printed circuit boards. You will be required to drill approximately 150 holes during the process of creating each robot in the book. An electric hand drill, like the one shown in **Figure 1.4**, can be used.

If you plan to build robots as a hobby, then a small drill press, like the one shown in **Figure 1.5**, would be a great idea. Using a drill press is highly recommended when drilling holes in printed circuit boards, where accuracy and straightness are important. These small drill presses don't cost much more than a good electric hand drill. I added an adjustable X-Y vise to the drill press in my work-

FIGURE 1.4

Hand held electric drill.

FIGURE 1.5

A small electric drill press with an X-Y adjustable vise.

FIGURE 1.6

Aluminum-cutting endmill.

shop. This makes it possible to mill aluminum if an endmill, like the one shown in **Figure 1.6**, is purchased from a machine shop supplier. The drill press can then double as a small milling machine.

You will need a set of drill bits like the ones pictured in **Figure 1.7**. The 5/32-inch and 1/4-inch drill bits are used most often during the projects. You will need to separately buy the small 1/32-inch and 3/64-inch bits that will be used to drill the component holes in the printed circuit boards.

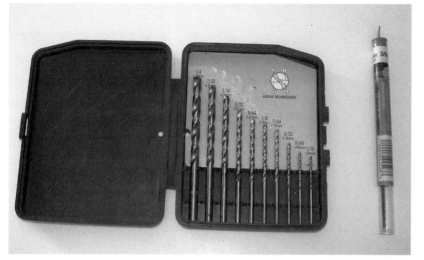

FIGURE 1.7

Drill bit set.

FIGURE 1.8

Various pliers, a
wrench, and
screwdrivers.

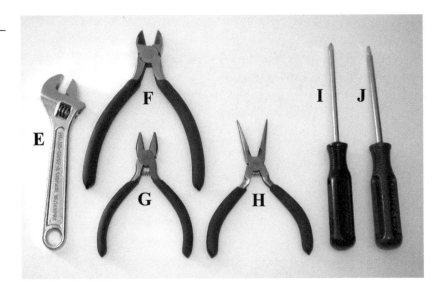

You will need an adjustable wrench (marked E in **Figure 1.8**), side
cutters (F), pliers (G), needle nose pliers (H), a Phillips screwdriver (I), and a Robertson screwdriver (J) during construction of the
robots. A set of miniature screwdrivers may be useful as well. The
needle nose pliers can be used to hold wire and small components in place while soldering, bending wire, and holding
machine screw nuts.

The wire strippers, shown in **Figure 1.9** (A), are used to strip the
protective insulation off wire, without cutting the wire itself. The
device is designed to accommodate a number of wire sizes you
will need. A pair of wire cutters (C) can cut wire when fabricating
jumper wires and wiring power to the circuits. You will need
rosin-core solder (B) when soldering components to the circuit
boards, creating jumper wires, and wiring the battery connectors
and power switches. To make soldering components to the printed circuit boards as easy as possible, buy the thinnest solder that
you can find. You will definitely need a chip-pulling tool (D) for
removing the PIC 16F84 chips from the 18-pin sockets. The PIC
16F84 will be inserted and removed from the sockets on the main
controller boards many times, as the software is changed and the

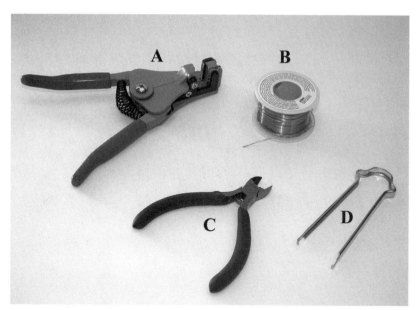

FIGURE 1.9

Wire strippers, cutters, solder, and a chip-pulling device.

PIC is reprogrammed during experiments. An adjustable work stand, like the one shown in **Figure 1.10** (M), will be useful when soldering components to circuit boards, or holding wires when soldering header connectors to the bare wires. A utility knife (N) will also be helpful when cutting heat-shrink tubing or small parts.

A soldering iron, similar to the one shown in **Figure 1.11**, will be required when building the main controller circuit boards and the sensor boards for each robot. An expensive soldering iron is not necessary, but the advantage to buying a good one is that the temperature can be set. A 15- to 25-watt pencil-style soldering iron will work and will help to protect delicate components from burning out.

An adjustable square (O) and a good ruler (P) will be required when measuring the cutting and drilling marks on the aluminum pieces that make up each robots' body and legs. You will need a hot glue gun (Q) and glue sticks at certain points in the construction. See **Figure 1.12**.

FIGURE 1.10

Adjustable work stand and utility knife.

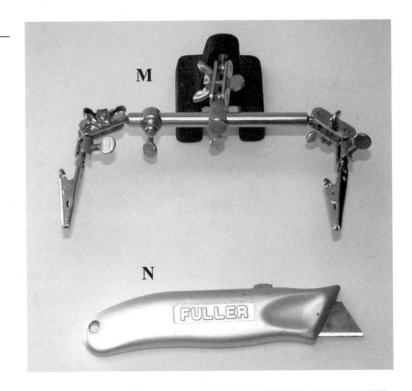

FIGURE 1.11

Soldering iron with adjustable temperature.

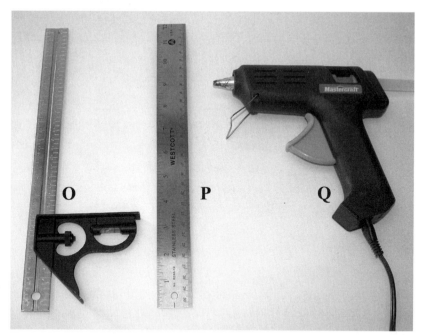

FIGURE 1.12

Adjustable square, ruler, and glue gun.

A hammer (R), shown in **Figure 1.13**, will be needed for bending aluminum, along with a metal file (S) to smooth the edges of metal pieces after they have been cut or drilled. You may use a tube of

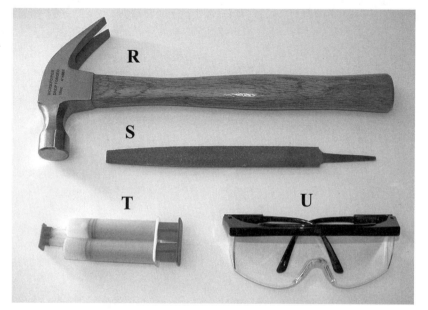

FIGURE 1.13

Hammer, file, epoxy, and safety glasses.

quick-setting epoxy (T) to secure parts. Safety glasses (U) should be worn at all times when cutting and drilling metal or soldering.

Test Equipment

To calibrate and troubleshoot the electronics, you will need a digital multimeter with frequency counting capabilities, similar to the Fluke 87 multimeter (**Figure 1.14**, left). When working with electronic circuits, a good multimeter is invaluable. The second multimeter in **Figure 1.14** (right) is manufactured by Circuit Test and measures capacitance, resistance, and inductance. It is nice to be able to measure the exact values of components when working on precise circuits, but in most cases, this is not necessary. If you are winding your own transformers or chokes, the ability to measure inductance will be helpful. The specific use of the multimeter will be explained during the construction of the robot's electronics in later chapters.

FIGURE 1.14

Fluke and Circuit Test multimeters.

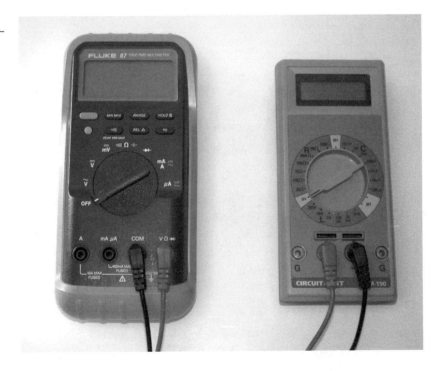

If you are really serious about electronics, then an oscilloscope, like the one pictured in **Figure 1.15**, is a great investment. This is the Tektronix TDS 210 dual channel, digital real-time oscilloscope, with a 60-MHz bandwidth. The TDS 210 on my bench also has the RS-232, GPIB, and centronics port module added, so that a hard copy of waveforms can be output. The great advantage to using an oscilloscope is the ability to visualize what is happening with a circuit. The new digital oscilloscopes also automatically calculate the frequency, period, mean, peak to peak, and true RMS of a waveform. You will probably need to use a regulated direct current (DC) power supply and a function generator quite often as well.

None of the equipment shown in **Figure 1.15** is required when building the robots in this book, but it will make your life as an

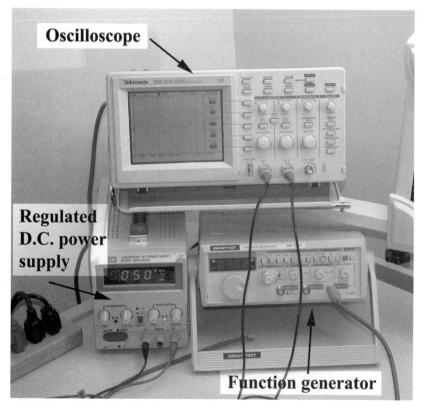

FIGURE 1.15

Oscilloscope, regulated DC power supply, and function generator.

electronics experimenter much easier. There is nothing more frustrating than finding out that a circuit you are working on is malfunctioning because of a dead battery or an oscillator calibrated to the wrong frequency. If you use a good power supply and oscilloscope when building and testing a circuit, the chance of these kinds of problems surfacing is much lower. I have always found that if I am working late at night and start to encounter a lot of small problems and make mistakes, the best thing to do is to shut my equipment down and get a good night's sleep. Sometimes the difference between frying an expensive chip or the circuit's working perfectly on the first try is just one misplaced component.

Construction Materials

The robots in this book are constructed using aluminum and fasteners that are readily available at most hardware stores. Five sizes of aluminum will be used. The first stock measures 1/2-inch wide by 1/8-inch thick, and is usually bought in lengths of 4 feet or longer. Many of the robot parts are constructed from aluminum, with the dimensions as shown in **Figure 1.16**.

FIGURE 1.16

1/2-inch by 1/8-inch aluminum stock.

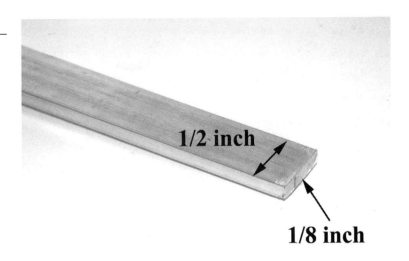

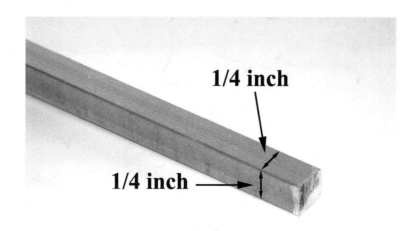

FIGURE 1.17

Aluminum stock with
1/4-inch by 1/4-inch
dimensions.

The second type of aluminum stock that will be used measures
1/4-inch × 1/4-inch, and is shown in **Figure 1.17**. It is usually
bought in lengths of 4 feet or longer as well.

The third kind of aluminum stock is 1/2-inch × 1/2-inch angle
aluminum, and is 1/16-inch thick, as shown in **Figure 1.18**.

The fourth type is 1/16-inch thick flat aluminum, as shown in
Figure 1.19, and it is usually bought in larger sheets. However,
most metal suppliers will cut it down for you. This thickness of
aluminum is great for cutting out custom parts and it is easy to

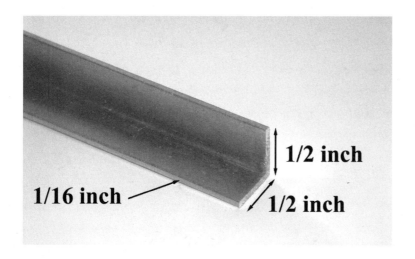

FIGURE 1.18

1/2-inch angle
aluminum.

FIGURE 1.19

1/16-inch thick flat aluminum.

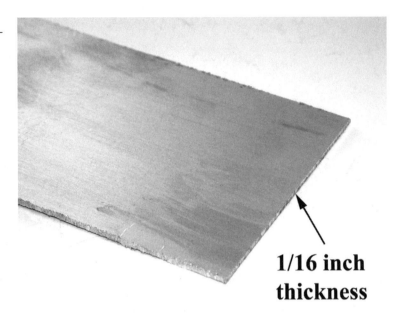

1/16 inch thickness

bend, making it ideal for the hobbyist experimenter. I buy all of my metal from a company called The Metal Supermarket (www.metalsupermarkets.com) because its prices are much lower than buying metal at a hardware store. Their friendly staff is always helpful, and will cut the stock to whatever size you require. I usually ask them to cut the raw stock in half so that it will fit into the back seat of my car.

The fifth type of stock that will be needed is 3/4-inch × 3/4-inch angle aluminum.

The fasteners that will be used are 6/32-inch diameter machine screws, nuts, lock washers, locking nuts, and nylon washers, as shown in **Figure 1.20**. Three different lengths of machine screws will be used: 1-inch, 3/4-inch, and 1/2-inch.

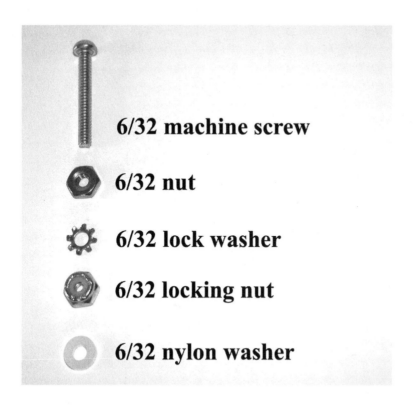

6/32 machine screw

6/32 nut

6/32 lock washer

6/32 locking nut

6/32 nylon washer

FIGURE 1.20

6/32-inch diameter machine screw, lock washer, nuts, and nylon washer.

Summary

Now that all the tools, test equipment, and materials necessary to build robots have been covered, you should have a good idea about what will be necessary to build the robots in this book. In the next chapter, the fabrication of printed circuit boards will be discussed so that you can make your own professional-looking boards.

2
Printed Circuit Board Fabrication

Four robot projects are in this book. Each robot will require a controller and sensor circuit boards. The most efficient way of implementing the circuit designs is to create printed circuit boards (PCBs). The great thing about each project is that the finished PCB artwork is included, along with a parts placement diagram. All of the circuit boards and robots in this book have been built and tested to ensure that they function as described. If you decide not to fabricate PCBs, most of the circuits are simple enough to construct on standard perforated circuit board (holes spaces 0.10-inch on centers) using point-to-point wiring if you wish. I don't recommend this method because one misplaced or omitted wire can cause hours of frustration.

The easiest way to produce quality PCBs is by using the positive photo fabrication process. To fabricate the PCBs for each robot project, photocopy the PCB artwork onto a transparency. Make sure that the photocopy is the exact size of the original. For convenience, you can download the artwork files for each robot project from the Thinkbotics Web site, located at www.thinkbotics.com, and print the file onto a transparency using a laser or ink-jet printer with a minimum resolution of 600 dpi. **Figure 2.1** shows the artwork for a

PCB artwork printed onto transparency film.

circuit board that has been printed onto transparency film using an ink-jet printer.

After successfully transferring the artwork to a transparency, the following instructions can be used to create a board. A 4- × 6-inch presensitized positive copper board is ideal for all of the projects presented in this book. When you place the transparency on the copper board, it should be oriented exactly as shown in each chapter. Make any sensor boards that go with the particular project at the same time. A company that specializes in providing presensitized copper boards and all the chemistry needed to fabricate boards is M.G. Chemicals. Information on how to obtain all of the supplies can be found on its Web site: www.mgchemicals.com. **Figure 2.2** shows the developer, ferric chloride, and presensitized copper board that will be used for fabricating the circuit boards.

FIGURE 2.2

Photo fabrication kit.

Follow the next six steps to make your own PCBs:

1. **Setup**—Protect surrounding areas from developer and other splashes that may cause etching damage. Plastic is ideal for this. Work under safe light conditions. A 40-W incandescent bulb works well. Do not work under fluorescent light. Just prior to exposure, remove the white protective film from the presensitized board. Peel it back carefully.

2. **Exposing your board**—For best results, use the M.G. Chemicals cat. #416-X exposure kit. However, any inexpensive lamp fixture that will hold two or more 18-inch fluorescent tubes is suitable.

   Directions: Place the presensitized board, copper side toward the exposure source. Positive film artwork should be laid onto the *presensitized* copper side of the board and positioned as desired. Artwork should have been produced by a 600-dpi or better printer. If you don't have a printer that can handle 600 dpi, then make two transparencies and lay them on top of each other. Make sure that the traces line up perfectly, and then staple them together. A glass weight should then be used to cover the artwork, ensuring that no light will pass under the traces (approximately 3-mm glass thickness or greater works best). Use a 10-minute exposure time at a distance of 5 inches.

3. **Developing your board**—The development process removes any photoresist that was exposed through the film positive to ultraviolet light. **Warning: The developer contains sodium hydroxide and is highly corrosive. Wear rubber gloves and eye protection while using it. Avoid contact with eyes and skin. Flush thoroughly with water for 15 minutes if it is splashed in eyes or on the skin.**

   Directions: Using rubber gloves and eye protection, dilute one part M.G. cat. #418 developer with 10 parts tepid water (weaker is better than stronger). In a plastic tray, immerse the board, copper side up, into the developer, and you will quickly see an image appear while you are lightly brushing the resist with a foam brush. This should be completed within one to two minutes. Immediately neutralize the development action by rinsing the board with water. The exposed resist must be removed from the board as soon as possible. When you are done with the developing stage, the only resist remaining will be covering what you want your circuit to be. The rest should be completely removed.

4. **Etching your board**—For best results, use the 416-E Professional Etching Process Kit or 416-ES Economy Etching Kit. The most popular etching matter is ferric chloride, M.G. cat. #415, an aqueous solution that dissolves most metals. **Warning: This solution is normally heated up during use, generating unpleasant and caustic vapors; adequate venti-**

lation is very important. Use only glass or plastic contain-
ers. Keep out of reach of children. May cause burns or
stain. Avoid contact with skin, eyes, or clothing. Store in
plastic container. Wear eye protection and rubber gloves.

If you use cold ferric chloride, it will take a long time to etch
the board. To speed up the etching process, heat up the solution. A simple way of doing this is to immerse the ferric chloride bottle or jug in hot water, adding or changing the water
to keep it heating. A thermostat-controlled crock pot is also
an effective way to heat ferric chloride, as are thermostati-
cally controlled submersible heaters—(glass enclosed, such
as an aquarium heater). An ideal etching temperature is 50°C
(120°F). Be careful not to overheat the ferric chloride. The
absolute maximum working temperature is about 57°C
(135°F). The warmer your etch solution, the faster your
boards will etch. Ferric chloride solution can be used over
and over again, until it becomes saturated with copper. As
the solution becomes more saturated, the etching time will
increase. Agitation assists in removing unwanted copper
faster. This can be accomplished by using air bubbles from
two aquarium air wands with an aquarium air pump. Do not
use an aquarium air stone. The etching process can be assist-
ed by brushing the unwanted resist with a foam brush while
the board is submerged in the ferric chloride. After the etching process is completed, wash the board thoroughly under
running water. Do not remove the remaining resist protecting
your circuit or image, as it protects the copper from oxidation. If you require it to be removed, use a solvent cleaner.
Figure 2.3 shows an etched board ready for drilling.

5. **Drilling and parts placement**—Use a 1/32-inch drill bit to
   drill all the component holes on the PCB. Drill the holes for
   larger components with a 3/64-inch bit where indicated. Drill
   any holes that will be used to mount the circuit board at this

FIGURE 2.3

An etched board ready for drilling.

time. It is best to use a small drill press, like the one shown in **Figure 2.4**, rather than a hand drill, when working with circuit boards. This is to ensure that the holes are drilled straight and accurately.

6. **Soldering your board**—Removal of resist is not necessary when soldering components to your board. When you leave the resist on, your circuit is protected from oxidation. Tin-plating your board is not necessary. In the soldering process, the heat disintegrates the resist underneath the solder, producing an excellent bond.

Summary

In the next chapter, the PIC microcontroller and how it is programmed will be described. Chapter 3 covers the use of compilers, hardware programmers, and the use of a development studio designed to speed up programming and debugging.

FIGURE 2.4

A small drill press used to drill holes in a PCB.

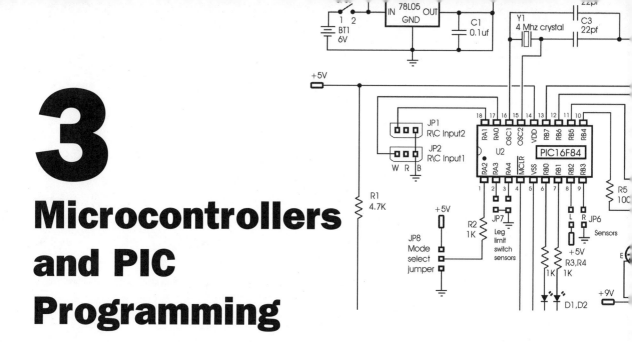

3

Microcontrollers and PIC Programming

Microcontrollers

The microcontroller is an entire computer on a single chip. The advantage of designing around a microcontroller is that a large amount of electronics needed for certain applications can be eliminated. This makes it the ideal device for use with mobile robots and other applications where computing power is needed. The microcontroller is popular because the chip can be reprogrammed easily to perform different functions, and is very inexpensive. The microcontroller contains all the basic components that make up a computer. It contains a central processing unit (CPU), read-only memory, random-access memory (RAM), arithmetic logic unit, input and output lines, timers, serial and parallel ports, digital-to-analog converters, and analog-to-digital converters. The scope of this book is to discuss the specifics of how the microcontroller can be used as the processor for the various robots that will be built.

PIC 16F84 MCU

Microchip technology has developed a line of reduced instruction set computer (RISC) microprocessors called the programmable interface controller (PIC). The PIC uses what is known as "Harvard architecture." Harvard uses two memories and separate busses. The first memory is used to store the program, and the other is to store data. The advantage of this design is that instructions can be fetched by the CPU at the same time that RAM is being accessed. This greatly speeds up execution time. The architecture commonly used for most computers today is known as Von Neumann architecture. This design uses the same memory for control and RAM storage, and slows down processing time.

We will be using the PIC 16F84, shown in **Figure 3.1**, as the processor for the robots in the book. This device can be reprogrammed over and over because it uses flash read-only memory for program storage. This makes it ideal for experimenting because the chip does not need to be erased with an ultraviolet light source every time you need to tweak the code or try something new.

The PIC 16F84 is an 18-pin device with an 8-bit data bus and registers. We will be using a 4-MHz crystal for the clock speed. This is very fast for our application when you consider that it is run-

FIGURE 3.1

Pinout of the PIC 16F84 microcontroller.

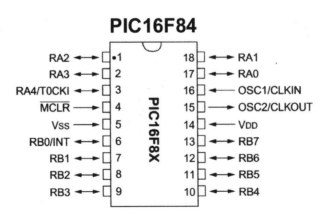

PIC16F84

RA2	1	18 RA1
RA3	2	17 RA0
RA4/T0CKI	3	16 OSC1/CLKIN
MCLR	4	15 OSC2/CLKOUT
Vss	5	14 VDD
RB0/INT	6	13 RB7
RB1	7	12 RB6
RB2	8	11 RB5
RB3	9	10 RB4

PIC16F8X

ning machine code at 4 million cycles per second. The PIC 16F84 is equipped with two input/output (I/O) ports, port A and port B. Each port has two registers associated with it. The first register is the TRIS (Tri State) register. The value loaded into this register determines if the individual pins of the port are treated as inputs or outputs. The other register is the address of the port itself. Once the ports have been configured using the TRIS register, data can then be written or read to the port using the port register address.

Port B has eight I/O lines available and Port A has five I/O lines. For example, the first robot project in the book details the construction and programming of a robotic frog. This project will use the same main controller circuit board as the hexapod robot featured in the book *Insectronics* so that readers who have built the Insectronic robot will be able to *jump* right into this project. The frog will be using all eight I/O lines of Port B and all five lines of Port A, as shown in **Figure 3.2**.

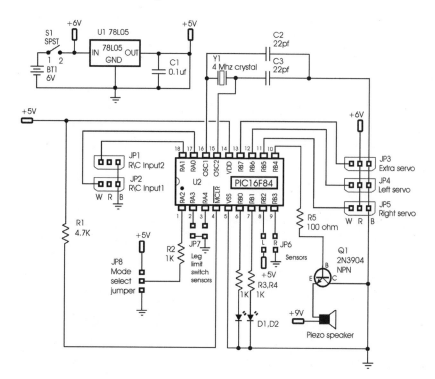

FIGURE 3.2

Frogbotics main controller board schematic.

Table 3.1 shows how the various pins of Port A and Port B will be used as inputs and outputs to control the different functions of the frog robot. It is useful to have a list of the various I/Os connected to the ports when programming.

TABLE 3.1	Port B	Configuration	Robot connection
PIC 16F84 Port A and B Connection Table	RB0	Output	Left light-emitting diode
	RB1	Output	Right light-emitting diode
	RB2	Input	Sensor input
	RB3	Input	Sensor input
	RB4	Output	Piezoelectric buzzer
	RB5	Output	Right servo
	RB6	Output	Left servo
	RB7	Output	Extra servo
	Port A	Configuration	Robot connection
	RA0	Input	Radio control input 1
	RA1	Input	Radio control input 2
	RA2	Input	Mode select jumper
	RA3	Input	Left leg limit switch
	RA4	Input	Right leg limit switch

PicBasic Pro Compiler

MicroEngineering Labs developed the PicBasic Pro Compiler, shown in Figure 3.3. It is a programming language that makes it quick and easy to program Microchip Technology's powerful PICmicro micro-

FIGURE 3.3

PicBasic Pro Compiler.

controllers. It can be purchased from microEngineering Labs, whose Web site is located at www.microengineeringlabs.com.

The BASIC language is much easier to read and write than Microchip assembly language, and will be used to program the robots in this book. The PicBasic Pro Compiler is "BASIC Stamp II-like," and has most of the libraries and functions of both the BASIC Stamp I and II. Because it is a true compiler, programs execute much faster, and may be longer than their Stamp equivalents.

One of the advantages of the PicBasic Pro Compiler is that it uses a real IF..THEN..ELSE..ENDIF, instead of the IF..THEN(GOTO) of the Stamps. These and other differences are spelled out in the PBP manual.

PicBasic Pro (PBP) defaults to create files that run on a PIC 16F84-04/P clocked at 4 MHz. Only a minimum of other parts are necessary: two 22pf capacitors for the 4-MHz crystal, a 4.7K pull-up resistor tied to the /MCLR pin, and a suitable 5-volt power supply. Many PICmicros other than the 16F84, as well as oscillators of frequencies other than 4 MHz, may be used with the PicBasic Pro Compiler.

The PicBasic Pro Compiler produces code that may be programmed into a wide variety of PICmicro microcontrollers having from 8 to 84 pins and various on-chip features, including A/D converters, hardware timers, and serial ports. For general purpose PICmicro development using the PicBasic Pro Compiler, the PIC 16F84, 16F876, and 16F877 are the current PICmicros of choice. These microcontrollers use flash technology to allow rapid erasing and reprogramming to speed program debugging. With the click of the mouse in the programming software, the flash PICmicro can be instantly erased and then reprogrammed again and again. Other PICmicros in the 12C67x, 14C000, 16C55x, 16C6xx, 16C7xx, 16C9xx, 17Cxxx, and 18Cxxx series are either *one-time programmable* (OTP) or have a quartz window in the top (JW) to allow erasure by exposure to ultraviolet light for several minutes. The PIC 16F84 and 16F87x devices also contain between 64 and 256 bytes of nonvolatile data memory that can be used to store program data and other parameters, even when the power is turned off. This data area can be accessed simply by using the PicBasic Pro Compiler's READ and WRITE commands. (Program code is always permanently stored in the PICmicro's code space, whether the power is on or off.)

By using a flash PICmicro for initial program testing, the debugging process may be sped along. Once the main routines of a program are operating satisfactorily, a PICmicro with more capabilities or expanded features of the compiler may be utilized.

Software Installation

The PicBasic Pro files are compressed into a self-extracting file on the diskette. They must be uncompressed to your hard drive before use. To uncompress the files, create a subdirectory on your hard drive called PBP or another name of your choosing by typing:

md PBP

at the DOS prompt. Change to the directory:

cd PBP

Assuming the distribution diskette is in drive a:, uncompress the files into the PBP subdirectory:

a:\pbpxxx -d

where xxx is the version number of the compiler on the disk. Don't forget the -d option on the end of the command. This ensures that the proper subdirectories within PBP are created.

Make sure that FILES and BUFFERS are set to at least 50 in your CONFIG.SYS file. Depending on how many FILES and BUFFERS are already in use by your system, allocating an even larger number may be necessary.

See the README.TXT file on the diskette for more information on uncompressing the files. Also, read the READ.ME file that is uncompressed to the PBP subdirectory on your hard drive for the latest PicBasic Pro Compiler information. **Table 3.2** lists the different PicBasic Pro Compiler statements that are available to the PICmicro software developer.

TABLE 3.2	Statement	Description
PicBasic Pro Statement Reference	@	Insert one line of assembly language code.
	ADCIN	Read on-chip analog to digital converter.
	ASM..ENDASM	Insert assembly language code section.
	BRANCH	Computed GOTO (equiv. to ON..GOTO).
	BRANCHL BRANCH	Out of page (long BRANCH).
	BUTTON	Debounce and auto-repeat input on specified pin.
	CALL	Call assembly language subroutine.
	CLEAR	Zero all variables.
	CLEARWDT	Clear (tickle) Watchdog Timer.
	COUNT	Count number of pulses on a pin.
	DATA	Define initial contents of on-chip EEPROM.
	DEBUG	Asynchronous serial output to fixed pin and baud.
	DEBUGIN	Asynchronous serial input from fixed pin and baud.
	DISABLE	Disable ON DEBUG and ON INTERRUPT processing.
	DISABLE DEBUG	Disable ON DEBUG processing.
	DISABLE INTERRUPT	Disable ON INTERRUPT processing.
	DTMFOUT	Produce touch-tones on a pin.
	EEPROM	Define initial contents of on-chip EEPROM.
	ENABLE	Enable ON DEBUG and ON INTERRUPT processing.
	ENABLE DEBUG	Enable ON DEBUG processing.

(continued on next page)

Statement	Description	TABLE 3.2
ENABLE INTERRUPT	Enable ON INTERRUPT processing.	PicBasic Pro Statement Reference (continued)
END FOR..NEXT	Stop execution and enter low power mode.	
FOR..NEXT	Repeatedly execute statements.	
FREQOUT	Produce up to 2 frequencies on a pin.	
GOSUB	Call BASIC subroutine at specified label.	
GOTO	Continue execution at specified label.	
HIGH	Make pin output high.	
HSERIN	Hardware asynchronous serial input.	
HSEROUT	Hardware asynchronous serial output.	
I2CREAD	Read bytes from I2C device.	
I2CWRITE	Write bytes to I2C device.	
IF..THEN..ELSE..ENDIF	Conditionally execute statements.	
INPUT	Make pin an input.	
LCDIN	Read from LCD RAM.	
LCDOUT	Display characters on LCD.	
{LET}	Assign result of an expression to a variable.	
LOOKDOWN	Search constant table for value.	
LOOKDOWN2	Search constant/variable table for value.	
LOOKUP	Fetch constant value from table.	
LOOKUP2	Fetch constant/variable value from table.	
LOW	Make pin output low.	
NAP	Power down processor for short period of time.	

(continued on next page)

33

	Statement	Description
TABLE 3.2 PicBasic Pro Statement Reference (continued)	ON DEBUG	Execute BASIC debug monitor.
	ON INTERRUPT	Execute BASIC subroutine on an interrupt.
	OUTPUT	Make pin an output.
	PAUSE	Delay (1mSec resolution).
	PAUSEUS	Delay (1uSec resolution).
	PEEK	Read byte from register. (Do not use.)
	POKE	Write byte to register. (Do not use.)
	POT	Read potentiometer on specified pin.
	PULSIN	Measure pulse width on a pin.
	PULSOUT	Generate pulse to a pin.
	PWM	Output pulse width modulated pulse train to pin.
	RANDOM	Generate pseudo-random number.
	RCTIME	Measure pulse width on a pin.
	READ	Read byte from on-chip EEPROM.
	READCODE	Read word from code memory
	RESUME	Continue execution after interrupt handling.
	RETURN	Continue at statement following last GOSUB.
	REVERSE	Make output pin an input or an input pin an output.
	SERIN	Asynchronous serial input (BS1 style).
	SERIN2	Asynchronous serial input (BS2 style).
	SEROUT	Asynchronous serial output (BS1 style).
	SEROUT2	Asynchronous serial output (BS2 style).

(continued on next page)

Statement	Description	
		TABLE 3.2
SHIFTIN	Synchronous serial input.	PicBasic Pro Statement Reference (continued)
SHIFTOUT	Synchronous serial output.	
SLEEP	Power down processor for a period of time.	
SOUND	Generate tone or white-noise on specified pin.	
SWAP	Exchange the values of two variables.	
TOGGLE	Make pin output and toggle state.	
WHILE..WEND	Execute statements while condition is true.	
WRITE	Write byte to on-chip EEPROM.	
WRITECODE	Write word to code memory.	
XIN	X-10 input.	
XOUT	X-10 output.	

Compiling A Program

For operation of the PicBasic Pro Compiler, you will need a text editor or word processor for creation of your program source file, some sort of PICmicro programmer such as the EPIC Plus Pocket PICmicro Programmer, and the PicBasic Pro Compiler itself. Of course you also need a PC to run it.

Follow this sequence of events:

First, create the BASIC source file for the program, using your favorite text editor or word processor. If you don't have a favorite, DOS EDIT (included with MS-DOS) or Windows NOTEPAD (included with Windows and Windows 95/98) may be substituted. A great text editor called Ultraedit is available at: www.ultraedit.com. It is geared towards the software developer and does not add any undesirable formatting characters that will cause the compiler to error out.

The source file name should (but is not required to) end with the extension .BAS. The text file that is created must be pure ASCII text. It must not contain any special codes that might be inserted by word processors for their own purposes. You are usually given the option of saving the file as pure DOS or ASCII text by most word processors.

Program 3.1 provides a good first test for programming a PIC and for testing the frog robot controller board when it is built in Chapter 4. You can type it in or download it from the author's Web site www.thinkbotics.com, and follow the links for book software.

The file is named frog-test.bas and is listed in **Program 3.1**. The BASIC source file should be created in or moved to the same directory where the PBP.EXE file is located.

PROGRAM 3.1

frog-test.bas program listing

```
'_____
' Name     : Frog-test.bas
' Compiler : PicBasic Pro MicroEngineering Labs
' Notes    : Program to test the main controller
'          : board by flashing LEDs, producing
'          : sounds and slowly rotating the servos
'_____
' set porta to inputs
trisa = %11111111
' set portb pins 2 & 3 to inputs
trisb = %00001100
'_____
' initialize variables
servo_pos_l   VAR BYTE
servo_pos_r   VAR BYTE
timer1        VAR BYTE
timer2        VAR BYTE
timer3        VAR BYTE
temp1         VAR BYTE
servo_r       VAR PORTB.5
servo_l       VAR PORTB.6
```

PROGRAM 3.1

frog-test.bas program
listing (continued)

```
switch_r        VAR PORTA.4
switch_l        VAR PORTA.3
led_l           VAR PORTB.1
led_r           VAR PORTB.0
piezo           VAR PORTB.4
'_____

low servo_l
low servo_r
start:
for temp1 = 1 to 10
    SOUND piezo, [80,4,100,2]
    low led_l
    low led_r
    pause 50
    high led_l
    high led_r
next temp1
SOUND piezo, [100,4,120,2,80,2,90,2]
low led_l
low led_r
rotate:
servo_pos_r = 170
gosub right_servo
servo_pos_l = 130
gosub left_servo
goto rotate
'_____


' subroutines to set servos
both_servo:
    for timer1 = 1 to 15
    pulsout servo_l,servo_pos_l
    pulsout servo_r,servo_pos_r
    pause 6
    next timer1
return
left_servo:
    for timer2 = 1 to 10
```

PROGRAM 3.1

frog-test.bas program
listing (continued)

```
        pulsout servo_l,servo_pos_l
        pause 6
        next timer2
    return
    right_servo
        for timer3 = 1 to 10
        pulsout servo_r,servo_pos_r
        pause 6
        next timer3
    return
    end
```

Once you are satisfied that the program you have written will work flawlessly, you can execute the PicBasic Pro Compiler by entering PBP, followed by the name of your text file at a DOS prompt. For example, if the text file you created is named frog-test.bas, at the DOS command prompt, enter:

PBP frog-test.bas

The compiler will display an initialization (copyright) message and process your file. If it likes your file, it will create an assembler source code file (in this case, named frog-test.asm) and automatically invoke its assembler to complete the task. If all goes well, the final PICmicro code file will be created (in this case, frog-text.hex). If you have made the compiler unhappy, it will issue a string of errors that will need to be corrected in your BASIC source file before you try compilation again.

To help ensure that your original file is flawless, it is best to start by writing and testing a short piece of your program, rather than writing an entire 100,000-line monolith all at once and then trying to debug it from end to end.

If you don't tell it otherwise, the PicBasic Pro Compiler defaults to creating code for the PIC 16F84. To compile code for PICmicros other than the F84, just use the -P command line option, described later in the manual, to specify a different target processor. For example, if you intend to run the above program, frog-test.bas, on a PIC 16C74, compile it using the command:

PBP -p16c74 frog-test.bas

An assembler source code file for frog-test.bas is also generated. It is called frog-test.asm. The assembler source code can be used as a guide if you want to explore assembly language programming because the listing shows the PicBasic Pro statement and the corresponding assembly code on the next line. The rest of the chapters discussing software will not be addressing assembly code. All we really need to be concerned with is the PicBasic source code and the generated .HEX machine code, as listed in **Program 3.2.**

If you do not have the resources to buy the PicBasic Pro compiler, simply type the listings of the .HEX files into a text editor and save the file with the program name and .HEX extension. All the program listings in the book can also be downloaded from www.thinkbotics.com to make things easier. However, I recommend buying a copy of the compiler if you wish to experiment, change, or customize the programs. If you decide to continue with robotics and electronics, you will eventually need to buy a compiler, such as PicBasic Pro, when working with microcontrollers.

```
:100000007B28A0003B200C080D04031976287020E3
:100010008413200880066400 0D280E288C0A03191A
:100020008D0F0B28800676288F0022088400200977
:100030003C2084138F0803197628 8F03091000E08B5
:1000400080389000F0309103031 9910003198F0359
:100050000031976282B283F2003010C1820088E1F37
:1000600020088E0803190301900F382880061F28E6s
:100070003928000022288FF3A8417800576280D08C9
```

PROGRAM 3.2

frog-test.hex program
listing (continued)

```
:100080000C0403198C0A80300C1A8D060C198D068D
:100090008C188D060D0D8C0D8D0D76288F018E0020
:1000A000FF308E07031C8F07031C762803308D005A
:1000B000DF305C2050288D01E83E8C008D09FC303B
:1000C000031C65288C07031862288C0764008D0FB9
:1000D00062280C186B288C1C6F2800006F28080001
:1000E0008C098D098C0A03198D0A080083130313E8
:1000F0008312640008008316FF3085000C308600F0
:10010000831206138316061383128612831686 1231
:100110008312 0130A60064000B3026020318B028B9
:10012000 0630A2001030A00050308E0004301420A1
:1001300064308E0002301420861083168610 8312DD
:100140000610831606 10323083124E208614831652
:10015000861083120614831606 10831 2A60F8B28AE
:100160000630A2001030A00064308E00043014204D
:1001700078308E0002301420 50308E00023014206F
:100180005A308E000230142086108316861083 1297
:1001900006108316061083122AA30A50000218230B3
:1001A000A400ED20CC280130A70064001030270205
:1001B0000318EC2824088C008D01063084004030A0
:1001C00012025088C008D0106308400203001209C
:1001D00006304E20A70FD52808000130A800640083
:1001E000B3028020318FF2824088C008D010630EC
:1001F00084004030012006304E20A80FEF28080070
:100200000130A90064000B3029020318122925 08C7
:100210008C008D0106308400203001200 6304E20F5
:0A022000A90F02290800630013294A
:02400E00F53F7C
:00000001FF
```

Using the EPIC Programmer to Program the PIC

The two steps left are putting your compiled program into the PICmicro microcontroller and testing it. The PicBasic Pro Compiler generates standard 8-bit Merged Intel HEX (.HEX) files that may be used with any PICmicro Programmer, including the EPIC Plus

FIGURE 3.4

EPIC Programmer by microEngineering Labs.

Pocket PICmicro Programmer, shown in **Figure 3.4**. PICmicros cannot be programmed with BASIC Stamp programming cables.

An example of how a PICmicro is programmed using the EPIC Programmer with the DOS programming software follows. If Windows 95/98/NT is available, using the Windows version of EPIC Programmer software is recommended.

Make sure there are no PICmicros installed in the EPIC Programmer programming socket or any attached adapters. Hook the EPIC Programmer to the PC parallel printer port using a DB25 male-to-DB25 female printer extension cable. Plug the AC adapter into the wall and then into the EPIC Programmer (or attach two fresh 9-volt batteries to the programmer and connect the "Batt ON" jumper). The light-emitting diode (LED) on the EPIC Programmer may be on or off at this point. Do not insert a PICmicro into the programming socket when the LED is on or before the programming software has been started.

Enter:

EPIC

at the DOS command prompt to start the programming software. The EPIC software should be run from a pure DOS session or from a full-screen DOS session under Windows or OS/2. (Running under Windows is discouraged. Windows [all varieties] alters the system timing and plays with the port when you are not looking, which may cause programming errors.)

The EPIC software will look around to find where the EPIC Programmer is attached and get it ready to program a PICmicro. If the EPIC Programmer is not found, check all the above connections and verify that there is not a PICmicro or any adapter connected to the programmer.

Typing:

EPIC /?

at the DOS command prompt will display a list of available options for the EPIC software.

Once the programming screen is displayed, use the mouse to click on **Open file** or press **Alt-O** on your keyboard. Use the mouse (or keyboard) to select frog-test.hex or any other file you would like to program into the PICmicro from the dialog box. The file will load and you should see a list of numbers in the window at the left.

This is your program in PICmicro code. At the right of the screen is a display of the configuration information that will be programmed into the PICmicro. Verify that it is correct before proceeding. In general, the oscillator should be set to XT for a 4-MHz crystal, and the Watchdog Timer should be set to ON for PicBasic Pro programs. Most important, Code Protect must be OFF when programming any windowed (JW) PICmicro. You may not be able to erase a windowed PICmicro that has been code protected. **Figure** 3.5 shows the EPIC MS-DOS interface.

Insert a PIC 16F84 into the programming socket and click on **Program** or press **Alt-P** on the keyboard. The PICmicro will first be checked to make sure it is blank, and then your code will be programmed into it. If the PICmicro is not blank and it is a flash device, you can simply choose to program over it without erasing first. Once the programming is complete and the LED is off, it is time to test your program.

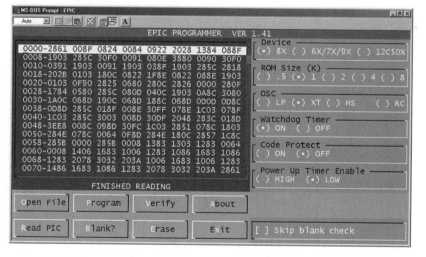

FIGURE 3.5

EPIC graphics user interface.

Testing the Controller Board

Later in Chapter 4, when the controller board is finished and the PIC 16F84 is programmed with the frog-test.hex program, insert the PIC into the socket on the controller board. Place the PIC into the 18-pin I.C. socket, with the notch and pin 1 facing toward the LEDs as shown in **Figure 3.6**.

Place four AA batteries in the 6-volt battery pack and secure it in position in the holder at the back of the robot. Make sure that the battery clip is attached, and then turn the power switch to the on position. If all is well, then the left and right LEDs should be alternatively flashing on and off, while the piezo element is producing robotic frog noises. When the flashing is finished, the servos should start rotating in a forward direction. This ensures that the 16F84 was programmed and that the controller board is functioning properly.

If nothing is happening when the power is switched on, try going through the process of programming the PIC again, and choose the verify option from the EPIC user interface. If the chip fails verifica-

FIGURE 3.6

PIC 16F84 inserted into I.C. socket on controller board.

tion, check the RS-232 cable and power supply to the programmer. If that does not work, try using a different 16F84 chip.

If there was no error when programming the PIC, insert it back into the controller board and make sure that pin 1 is facing toward the LEDs. Check the battery wiring and verify that the 6-V DC polarity is not reversed to the power connectors. Check the controller board for any missed components or cold solder connections.

MicroCode Studio Visual Integrated Development Environment

Mecanique's MicroCode Studio is a powerful, visual Integrated Development Environment (IDE), with an In Circuit Debugging (ICD) capability designed specifically for microEngineering Labs' PICBasic Pro Compiler. The MicroCode Studio user interface is shown in **Figure** 3.7.

This studio makes programming PIC microcontrollers very easy with a one-button process of compiling, assembling, and programming. MicroCode Studio is completely free for noncommercial use and can be downloaded at www.mecanique.co.uk/code-studio/. It is not time-limited in any way, and does not have any nag screens. However, you can only use one ICD model with MicroCode Studio. MicroCode Studio is not copyright-free. If you wish to redistribute MicroCode Studio, or make it available on another server, you must contact Mecanique and obtain permission first.

The main editor provides full syntax highlighting of your code, with context-sensitive keyword help and syntax hints. The code explorer allows you to automatically jump to include files, defines, constants, variables, aliases and modifiers, symbols, and labels that are contained within your source code. Full cut, copy, paste, and undo is provided, together with search and replace features. It also gives you the ability to identify and correct com-

FIGURE 3.7

MicroCode Studio makes PIC programming easy.

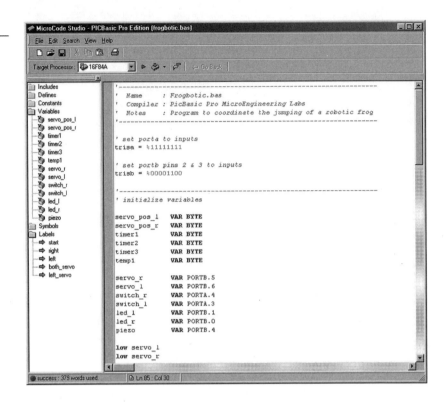

pilation and assembler errors. MicroCode Studio lets you view serial output from your microcontroller. It includes keyword-based context-sensitive help, and also supports MPASM and MPLAB.

It is easy to set up your compiler, assembler, and programmer options, or you can let MicroCode Studio do it for you with its built-in autosearch feature, as shown in **Figure 3.8**.

MicroCode Studio has support for MPLAB-dependent programmers such as PICStart Plus. Compilation and assembler errors can be easily identified and corrected using the error results window. Just click on a compilation error and MicroCode Studio will automatically take you to the error line. MicroCode Studio even comes with a serial communications window, allowing you to debug and view serial output from your microcontroller.

FIGURE 3.8

Automatically setting up the compiler.

With MicroCode Studio, you can start your preferred programming software from within the IDE. This enables you to compile and then program your microcontroller with just a few mouse clicks (or keyboard strokes, if you prefer). MicroCode Studio also supports MPLAB dependant programmers.

Using a Programmer with MicroCode Studio

The first thing you need to do is tell MicroCode Studio which programmer you are using.

Select VIEW...OPTIONS from the main menu bar, then select the PROGRAMMER tab, as shown in **Figure 3.9**. Next, select the Add New Programmer button. This will open the Add New Programmer wizard.

Select the programmer you want MicroCode Studio to use, then choose the Next button. MicroCode Studio will now search your

FIGURE 3.9

Adding a new
programmer.

computer until it locates the required executable. If your device uses MPLAB, you will be presented with two further screens, the select options and development mode screens. If your programmer is not in the list, you will need to create a custom programmer entry. Your programmer is now ready for use. When you press the Compile and Program button on the main toolbar, your PICBasic program is compiled and the programmer software is started. The hex filename and target device is automatically set in the programming software (if this feature is supported), ready for you to program the microcontroller, as shown in **Figure 3.10**.

MicroCode Studio in Circuit Debugger

The MicroCode Studio ICD enables you to execute a PICBasic Program on a host PIC microcontroller and view variable values, Special Function Registers (SFR), memory, and EEPROM as the program is running. Each line of source code is animated in the

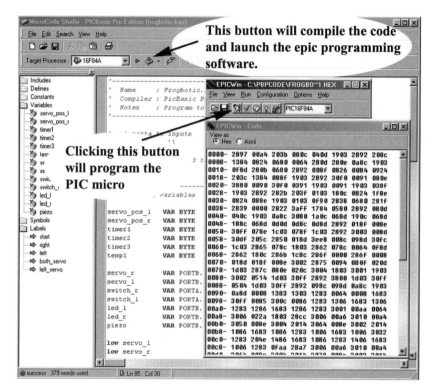

FIGURE 3.10

One button compile and programming using MicroCode Studio.

main editor window, showing you which program line is currently being executed by the host microcontroller. You can even toggle multiple breakpoints and step through your PICBasic code line by line.

Using the MicroCode Studio ICD can really accelerate program development. It's also a lot of fun and a great tool for learning more about programming PIC microcontrollers.

Summary

Now that the concept of programming and compiling code for microcontrollers has been covered, it will be easy to program the robots in the following chapters. Using MicroCode Studio for creating your source code, compiling the code, and programming PIC microcontrollers makes development much faster.

4
Frogbotic:
Build Your Own
Robotic Frog

Frogs and Toads

There are more than 4,100 species of frogs and toads, making them the largest group of amphibians. The majority lives in tropical environments, mostly in or close to fresh water. In adulthood, frogs and toads are characterized by the absence of a tail. The frog's hind limbs are much larger than their front limbs, enabling them to jump very long distances.

There is much diversity among frogs and toads. There are species that use their legs to swim, burrow into the soil, climb trees, and glide through the air, in addition to jumping and crawling. The primary senses of frogs and toads are vision and hearing. Many frogs and toads use loud calls to communicate with one another. Frogs and toads typically lay their eggs in water. The eggs hatch into larvae (tadpoles), which have spherical bodies and are herbivorous. Adult frogs and toads are carnivorous, feeding mostly on insects. They are generally only active at night.

The biologically inspired robot in this chapter is based on the frog and its capability to achieve locomotion by jumping. This locomo-

FIGURE 4.1

A tree frog and its biologically inspired robotic counterpart.

tion is achieved by releasing the energy stored in the frog's hind legs. **Figure 4.1** shows a tree frog, along with its biologically inspired mechanical counterpart.

Overview of the Frogbotic Project

The robotic frog to be built possesses two spring-loaded hind legs that are used to achieve locomotion by jumping, as shown in **Figures 4.2** and **4.3**. The functions of the leg mechanisms, sensors, and leg position limit switches are controlled by a Microchip PIC 16F84 microcontroller.

The spring of each leg is independently loaded with a mechanism that uses a standard servo, modified for continuous rotation. A close-up of the spring-loading mechanism is shown in **Figure 4.4**. When the servo is rotated to the position where the cam-like device is fully set and the spring is loaded, a limit switch is triggered. At this point, the microcontroller stops the servo and holds this position until both legs are in jumping position.

FIGURE 4.2

Robot frog leg
mechanism—outside
view.

FIGURE 4.3

Robot frog leg
mechanism—inside
view.

FIGURE 4.4

Spring-loading
mechanism with limit
switch sensor.

When both servos have been positioned so that the springs are
loaded and the legs are in their jumping position, the microcontroller gives both servos the command to move forward. This
moves the lever past the position where the spring is loaded, at
which time the spring quickly pulls the upper leg mechanism
downward, giving the legs enough energy to leap the frog forward.

R/C Servo Motors

The R/C servo is a geared, direct current motor with a built-in
positional feedback control circuit, as pictured in **Figure 4.6**. This
makes it ideal for use with small robots because the experimenter
does not have to worry about motor control electronics.

A *potentiometer* is attached to the shaft of the motor and rotates
along with it. For each position of the motor shaft and potentiometer, a unique voltage is produced. The input control signal is
a variable-width pulse between 1 and 2 milliseconds (ms), deliv-
ered at a frequency between 50 and 60 Hz, which the servo internally converts to a corresponding voltage. The servo feedback cir-

cuit constantly compares the potentiometer signal to the input control signal provided by the microcontroller. The internal comparator moves the motor shaft and potentiometer either forward or in reverse, until the two signals are the same. Because of the feedback control circuit, the rotor can be accurately positioned and will maintain the position as long as the input control signal is applied. The shaft of the motor can be positioned through 180 degrees of rotation, depending on the width of the input signal.

The PicBasic Pro language makes servo control with a PIC microcontroller easy, using a command called Pulsout. The syntax is Pulsout Pin, Period. A pulse is generated on Pin of specified Period. Toggling the pin twice generates the pulse; thus, the initial state of the pin determines the polarity of the pulse. Pin is automatically made an output. Pin may be a constant, 0–15, or a variable that contains a number between 0 and 15 (e.g., B0) or a pin name (e.g., PORTA.0).

The resolution of Pulsout is dependent on the oscillator frequency. Since we are using a 4-MHz oscillator, the Period of the generated pulse will be in 10 microsecond increments. To send a pulse to port B on pin 7 that is 1.4 ms long (at 4 MHz, 10 µs × 140 = 1400 µs or 1.4 ms), the command would be: Pulsout PortB.7,140. To illustrate the kind of signal being produced by the microcontroller, see **Figure 4.5**. The oscilloscope trace for channel 1 was generated with the Pulsout command configured to produce a 1.4-ms pulse at 55.68 Hz, and the trace for channel 2 was configured for a 6-ms pulse, also at 55.68 Hz.

Modifying Servos for Continuous Rotation

The robot frog will use two standard R/C servos, modified for continuous rotation. This is because servos are inexpensive, can be controlled directly from a microcontroller, and will provide the torque needed to load the spring-driven jumping leg mechanisms.

FIGURE 4.5

Oscilloscope display of a 1.4-ms and 6.0-ms pulse train.

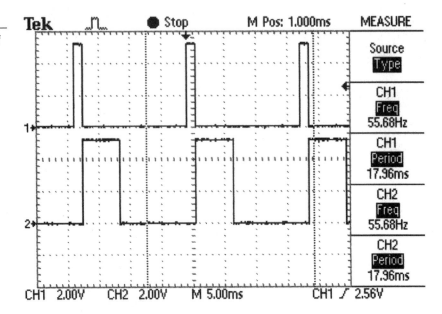

An unmodified servo has a rotational radius limited to approximately 180 degrees. For our application, we will need a full 360 degrees of continuous rotation. This is accomplished by taking the servo apart, removing a mechanical stop as well as removing the potentiometer, and replacing it with a fixed resistor network. Note that there may be differences between servos built by different manufacturers. The concept for modifying servos is basically the same for all servo types. Depending on the make, you may have to improvise and stray from the procedure a little. The servo in this example is a JR NES-527. The parts needed for this procedure are listed in **Table 4.1**.

TABLE 4.1

Parts Need for Modifying the Servos

Part	Quantity	Description
Resistors	4	2.4-KΩ resistors, 1/4-watt
Heat-shrink tubing	3 inches	Heat-shrink tubing

The instructions for modifying a standard servo are as follows:

1. Place the servo on a table and remove the servo horn and screw, as shown in **Figure 4.6**, if there is one attached.

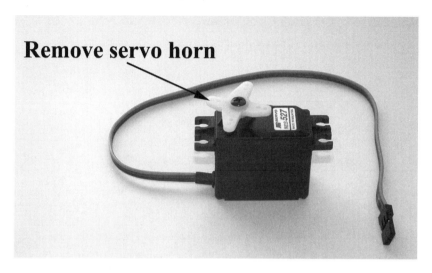

FIGURE 4.6

Remove servo horn and screw.

2. Flip the servo over so that the bottom is facing upward, and remove the four screws that hold the cover on. See **Figure 4.7** for details.

FIGURE 4.7

Servo cover screws removed from servo.

3. Next, turn the servo back over so that it is upright. Remove the top cover and the gears, as shown in **Figure 4.8**. You may need to use a small screwdriver to carefully break the seal and pry the cover off. When the cover has been removed, remove each of the gears in order, and place them somewhere safe. Leave the gear that connects to the motor in place.

FIGURE 4.8

Servo with cover and gears removed.

4. Now that the top cover is removed, open the bottom cover. Again, you may have to use a small screwdriver to get it open. Locate the small potentiometer and pry back the plastic clips that hold it in place, as shown in **Figure 4.9**. You may have to actually break them off. If you do break them off, make sure that they are removed from inside the servo and discarded. If glue is holding the potentiometer in place, scrape it off with a small screwdriver.

FIGURE 4.9

Potentiometer clips.

5. Turn the servo over so that the top is facing upward. Use a screwdriver to force the potentiometer shaft through the hole that it is mounted in, as shown in **Figure 4.10**. Pull the potentiometer all the way through and remove any glue holding the wires in place so that it resembles **Figure 4.11**.

FIGURE 4.10

Push potentiometer through mounting hole.

Push shaft through hole

FIGURE 4.11

Potentiometer removed from servo housing.

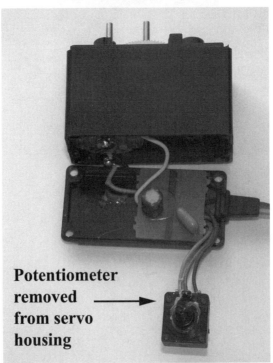

Potentiometer removed from servo housing

6. Use a soldering iron to de-solder the three wires that are attached to its leads, as shown in **Figure 4.12**. Take note of which wire is attached to the center terminal. Either mark the wire or write down the color, as this wire must be connected to the middle lead of the resistor network that will be fabricated in the next step.

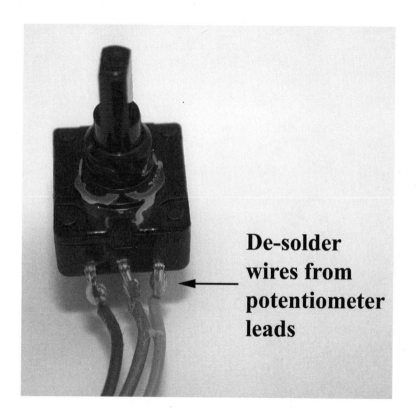

FIGURE 4.12

Potentiometer with wires attached.

7. For this step, you will need two 1/4-watt, 2.4 KΩ resistors to create a resistor network that will replace the potentiometer that was just removed. Try to select two resistors that have very close resistance values, although it is not extremely important, since any discrepancies can be compensated for in the control software. Cut the resistor leads to a length of 3/8-inch. Twist two of the ends together and solder, as shown in **Figure 4.13**.

FIGURE 4.13

Resistor network.

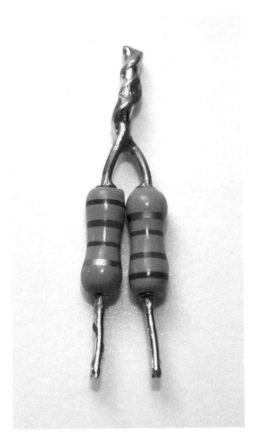

8. Cut three pieces of heat-shrink tubing, and slip each one over each of the wires that were attached to the potentiometer. Solder the middle wire from the potentiometer to the two resistor leads that are twisted together. Solder the left wire to the left resistor lead, and solder the right wire to the right resistor lead of the resistor network. Push the heat-shrink tubing up over the solder connections and shrink into place with a heat source. The finished resistor network with the wires soldered and the heat-shrink tubing in place should look like the one in **Figure 4.14**. Once this is complete, push the resistor network and the wires back into the servo in the space where the potentiometer was previously.

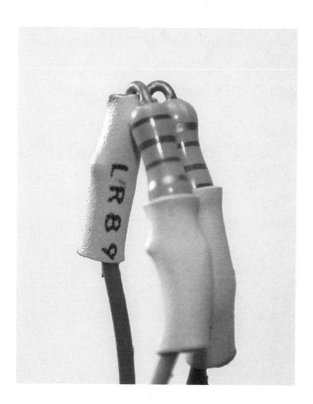

FIGURE 4.14

Resistor network with wires soldered into place.

9. Take the large output gear and locate the nub on its bottom side. Use a pair of side cutters to remove the nub, as shown in **Figure** 4.15. Use a file or a sharp knife to remove any excess plastic so that the bottom of the gear where the nub used to be is flat.

FIGURE 4.15

Removing the nub from the output gear.

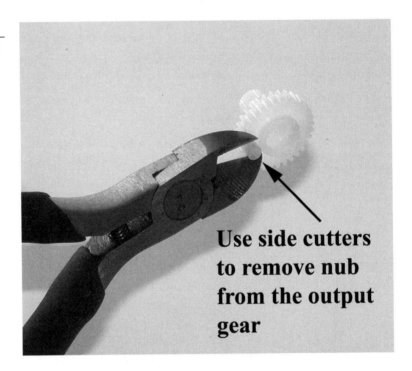

Use side cutters to remove nub from the output gear

10. Now that the gear has been modified, make sure that the bottom servo cover is in position. Replace the gears in the same order that they were removed from the servo. Use **Figure 4.16** as a guide when replacing the gears.

FIGURE 4.16

Servo gear placement.

11. Finally replace the top servo cover and secure it in place with the four screws that were removed during step 2. When the cover and screws are replaced, the servo should resemble the one shown in **Figure 4.17**. Be sure to mark the servo, indicating that it has been modified, since it will look exactly the same as an unmodified servo.

FIGURE 4.17

A servo modified for continuous rotation.

Controlling a Modified Servo

A modified servo is controlled in the same way as an unmodified servo. The only difference is that when the pulse width signal is sent to the servo, it will start turning the motor in the required direction and will continue to rotate as long as the signal is applied. Since the potentiometer that keeps track of the output gear position has been removed and replaced with the resistor network, the internal circuitry will think that the motor has not reached the specified position and will continue to seek for it in one direction or another. With identical resistors in the network, if a pulse with a width of 150 ms is sent to the servo, it will remain motionless. Since no two resistors are exactly the same, you may have to experiment with the pulse width value needed for the servo to remain motionless. It will probably be within the range of 147–153 ms. **Figure 4.18** illustrates how the modified servo will behave when control signals between 100 and 200 ms are applied. When a signal with a pulse width of 100 ms is applied to the servo's control line, the servo will move in a counterclockwise direction at full speed. The servo speed can be controlled by varying the pulse-width value, with 100 ms being the fastest speed in the counterclockwise direction, and 149 ms being the slowest. The same holds true for the servo rotating in

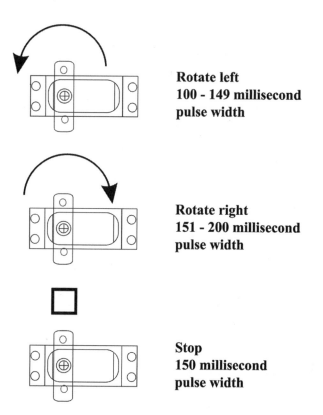

**Rotate left
100 - 149 millisecond
pulse width**

**Rotate right
151 - 200 millisecond
pulse width**

**Stop
150 millisecond
pulse width**

FIGURE 4.18

Pulse width values used to control a modified servo.

the clockwise direction, with 151 ms being the slowest speed and 200 ms being the fastest.

Mechanical Construction of Frogbotic

The construction of the robot frog will begin with the robot's body. The parts needed for the mechanical construction are listed in Table 4.2.

TABLE 4.2

Parts List for Frog Robot Mechanical Construction

Parts	Quantity
1/2-inch × 1/8-inch aluminum stock	4 feet, 2 inches
1/16-inch thick aluminum stock	12-inch × 12-inch piece
1/4-inch × 1/4-inch aluminum stock	2 inches
1/4-inch diameter nylon feet	2
6/32 × 1/2-inch machine screws	39
6/32 × 3/4-inch machine screws	8
6/32 nuts	23
6/32 lock washers	23
6/32 locking nuts	16
Standard R/C servo–modified	2
3/8-inch diameter × 5/8-inch spring	2

The body is constructed using a piece of 1/16-inch thick aluminum cut to a size of 4 × 7 inches. Use **Figure 4.19** as a guide to cutting and bending the aluminum piece. When the piece has been cut, use a file to remove any rough edges. Use **Figure 4.20** to measure and mark where all of the holes are to be drilled. Drill each of the holes with a 5/32-inch drill bit, except for the holes marked as being drilled with 1/4-inch and 7/64-inch bits. **Figure 4.21** shows the finished frog robot body on which all of the other mechanical and electronic components will be mounted.

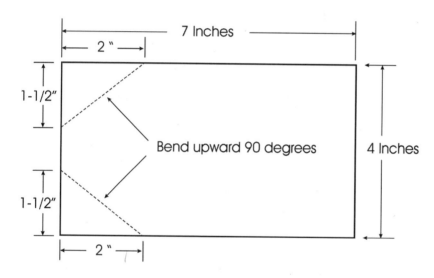

FIGURE 4.19

Cutting and bending guide for the Frogbotic's body.

FIGURE 4.20

Drilling guide for the
Frogbotic's body.

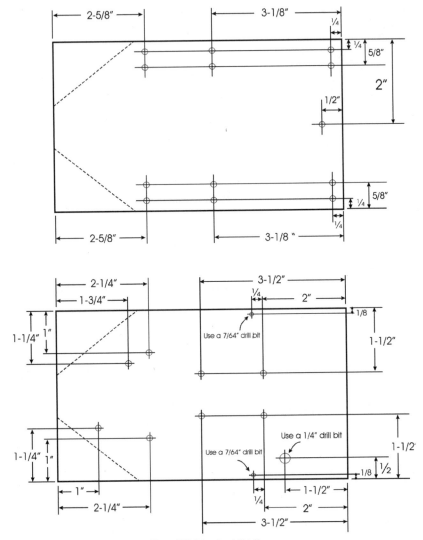

Holes are drilled with a 5/32 inch drill bit
except where marked.

FIGURE 4.21

Cut, bent, and drilled aluminum for the Frogbotic's body.

The next step is to fabricate six mounting brackets that will be used to attach the robot legs to the body and to fasten two leg sensor limit switches. Use **Figure 4.22** as a cutting and drilling guide to fabricate pieces A, B, C, D, E, and F out of 1/16-inch thick aluminum. Pieces A and B measure 1-3/4 inches in length, pieces C and D are 1-1/2 inches in length, and pieces E and F are 2-1/4 inches in length. Drill the holes with a 5/32-inch drill bit where indicated in **Figure 4.22**. The finished pieces are shown in **Figure 4.23**.

FIGURE 4.22

Cutting and drilling
guide for mounting
brackets.

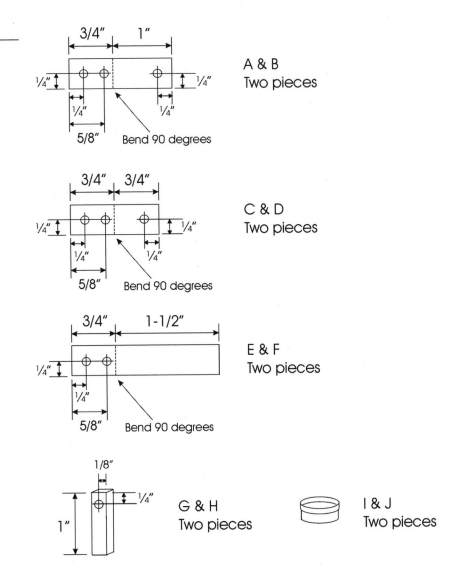

3/4"   1"

A & B
Two pieces

1/4"   1/4"

1/4"   1/4"

5/8"   Bend 90 degrees

3/4"   3/4"

C & D
Two pieces

1/4"   1/4"

1/4"   1/4"

5/8"   Bend 90 degrees

3/4"   1-1/2"

E & F
Two pieces

1/4"

1/4"

5/8"   Bend 90 degrees

1/8"

1/4"

G & H
Two pieces

I & J
Two pieces

1"

Holes are drilled with a 5/32 inch drill bit

FIGURE 4.23

Finished mounting brackets.

Fasten the leg mounting brackets and limit switch brackets to the frog's body piece, as shown in **Figure 4.24**. Fasten each bracket in place using two 6/32-inch × 1/2-inch machine screws, lock washers, and nuts. The frog's body with the brackets mounted in position should look like the one in **Figure 4.24**.

Cut two pieces of the 1/4-inch × 1/4-inch aluminum, marked as G and H, to a length of 1 inch, and drill according to **Figure 4.22**.

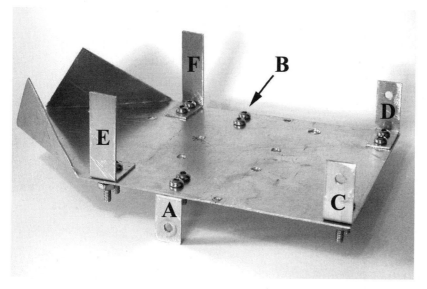

FIGURE 4.24

Mounting brackets fastened to the body.

FIGURE 4.25

Cut and drilled leg stops.

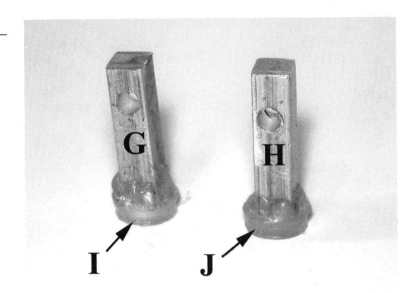

Use hot glue to fasten one of the 1/4-inch diameter plastic feet, marked as I in **Figure 4.22**, to the end of piece G. Do the same for pieces H and J. The finished leg stops are shown in **Figure 4.25**, and will be used to stop the legs from overtravelling when assembled later.

Using the 1/2-inch aluminum stock, cut and drill 10 pieces labeled K, L, M, N, O, P, Q, R, S, and T, as shown in **Figure 4.26**. Cut two pieces of 1/16-inch aluminum to a size of 1-1/2 inches × 2 inches. Photocopy the image in **Figure 4.27** onto a sheet of paper and use the enlarge feature until the dotted outline is exactly 1-1/2 inches × 2 inches. Another method is to scan the image into your computer and use a graphics editor program to make the enlargement and then print the image. Cut the images out and glue them to the aluminum pieces. Use a metal cutting band saw or a hack saw to cut the aluminum along the guide lines. Once the cuts have been made, bend the top part of the pieces upward, along the dotted lines, on 90-degree angles, as shown in **Figure 4.29**. These two pieces are the frog's feet and will be attached to pieces S and T.

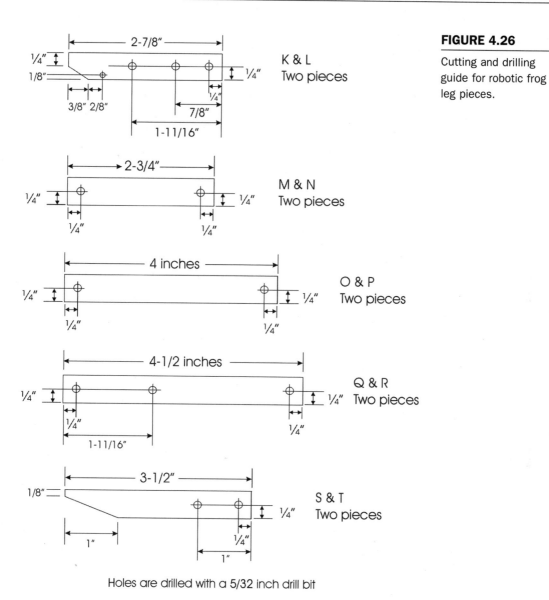

FIGURE 4.26

Cutting and drilling guide for robotic frog leg pieces.

Holes are drilled with a 5/32 inch drill bit

FIGURE 4.27

Cutting guide for robotic frog's feet.

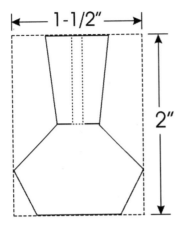

U & V
Two pieces

Attach the assembled leg stops (pieces G and H) to pieces L and K using two 6/32-inch × 3/4-inch machine screws, lock washers, and nuts, as shown in **Figure 4.28**. These assemblies will be part of the robot's leg-jumping mechanism.

Use hot glue to attach the robot's feet pieces U and V to pieces S and T on the sloped ends. **Figure 4.29** shows the feet pieces, U and V, attached to lower leg pieces S and T.

FIGURE 4.28

Assembled leg stops.

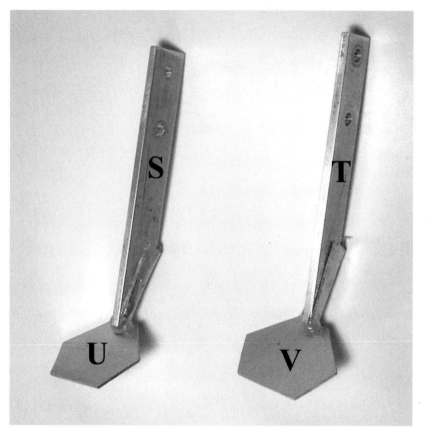

FIGURE 4.29

Feet attached to lower leg pieces S and T.

Assembling the Legs

Now that all of the individual leg pieces have been fabricated, it is time to assemble the legs. Starting with the frog's right leg, refer to **Figure 4.34** for overall parts placement. Place the part labeled L on a table and place a nylon washer over the 5/32-inch drill hole at the sloped end of the piece. Place the part labeled N on top and place another nylon washer on top of part N, lining up the holes. Next, place the part labeled P on top of the washer and insert a 6/32-inch × 3/4-inch machine screw through all three pieces and the nylon washers. **Figure 4.30** is an exploded view, illustrating how the parts are assembled. The nylon washers separating parts L, N, and P act as bearings. Secure in place with a 6/32-inch lock-

FIGURE 4.30

Exploded-view illustration of nylon washer bearing assembly.

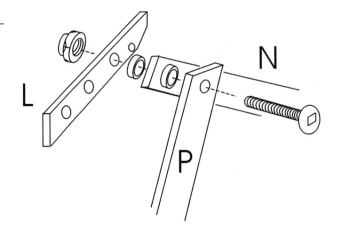

ing nut. Tighten the nut with enough torque to hold the parts in place, but allowing them to move freely. **Figure 4.31** shows the assembled parts. Take pieces R and T and assemble with piece R underneath T, placing a nylon washer between the two pieces, as shown in **Figure 4.32**.

FIGURE 4.31

Right leg subassembly made up of parts L, N, and P.

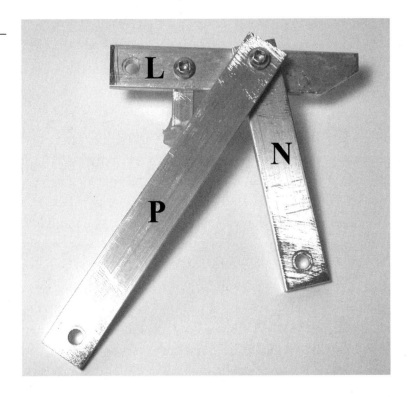

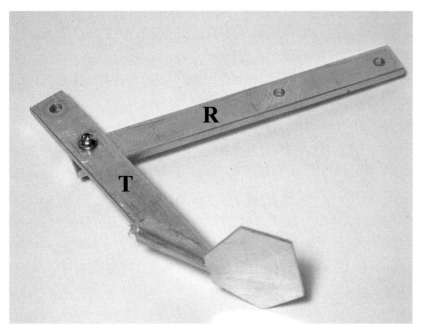

FIGURE 4.32

Lower right leg assembly made up of parts R and T.

To complete the right leg, take the subassembly made up of pieces L, N, and P and place it on top of the subassembly made up of pieces R and T, with a nylon washer between each of the holes. Secure in place with two 6/32-inch × 1/2-inch machine screws and locking nuts. Tighten the nuts with enough torque to hold the parts in place, but allowing them to move freely. Refer to **Figure 4.33** to see what the finished leg should look like.

FIGURE 4.33

Assembled right leg.

To assemble the left leg, refer to **Figure 4.34** for overall parts placement. Place the part labeled K on a table and place a nylon washer over the drill hole at the sloped end of the piece. Place the part labeled M on top and place another nylon washer on top of part M, lining up the holes. Next, place the part labeled O on top of the washer and insert a 6/32-inch × 3/4-inch machine screw through all three pieces and the nylon washers. The nylon washers separating parts K, M, and O act as bearings. Secure in place with a 6/32-inch locking nut. Tighten the nut with enough torque to hold the parts in place, but allowing them to move freely. Take pieces Q and S and assemble with piece Q underneath S, placing a nylon washer between the two pieces.

To complete the right leg, take the subassembly made up of pieces K, M, and O and place it on top of the subassembly made up of pieces Q and S, with a nylon washer between each of the holes. Secure in place with two 6/32-inch × 1/2-inch machine screws

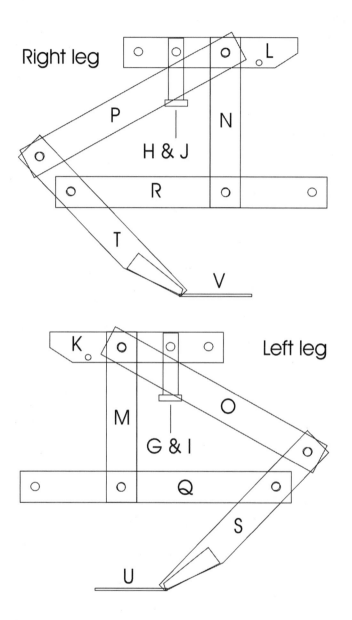

Right leg

Left leg

FIGURE 4.34

Parts placement
diagram for right and
left legs.

and locking nuts. Tighten the nuts with enough torque to hold the parts in place, but allowing them to move freely. The left leg is identical to the right leg, with the only difference being that the parts placement is a mirror of the right leg.

Attaching the Legs to the Robot's Body

Now that both the right and left legs have been constructed, it is time to attach them to the robot's body. Starting with the right leg, attach leg piece R to body mounting bracket B, and leg piece L to body mounting bracket D, with two 6/32-inch × 1/2-inch machine screws and locking nuts with nylon washers separating each piece, as shown in **Figure 4.35**. Tighten the nuts with enough torque to hold the parts in place, but allowing them to move freely.

FIGURE 4.35

Right leg attached to the robot's mounting brackets.

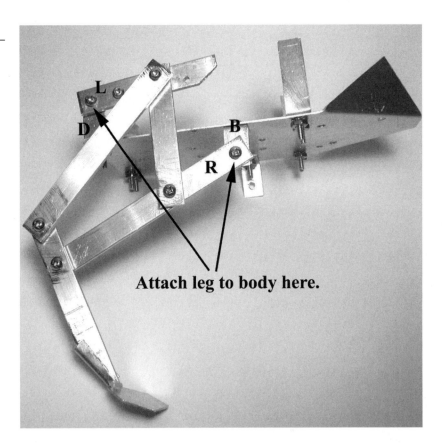

Attach leg to body here.

Take the left leg and attach leg piece Q to body mounting bracket A, and leg piece K to body mounting bracket C, with two 6/32-inch × 1/2-inch machine screws and locking nuts with nylon washers separating each piece. Tighten the nuts with enough torque to hold the parts in place, but allowing them to move freely. Refer to **Figure 4.24** and **Figure 4.34** for identification of parts. **Figure 4.36** shows the left and right legs attached to the mounting brackets.

The next step is to attach the leg springs. Cut two springs with a diameter of 3/8-inch to a length of 5/8-inch, like the one shown in **Figure 4.37**. Attach one end of each of the springs to leg pieces L and K, and the other ends to the robot's body, as shown in **Figure 4.38**. Make sure that the springs fit snugly so that they do not fall loose when the legs are retracted.

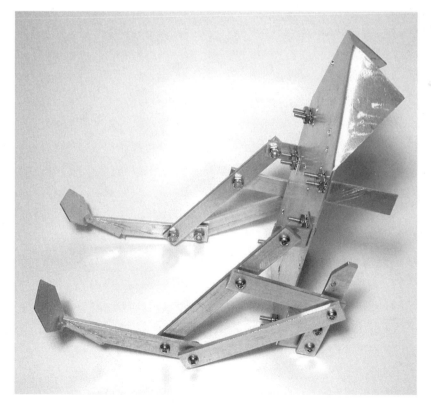

FIGURE 4.36

Right and left legs attached to the robot body.

FIGURE 4.37

Spring used for the right and left legs.

FIGURE 4.38

Spring attached to the robot's leg and body.

Spring

Fabricating the Servo Mounts

Use 1/16-inch thick aluminum to create two servo mounts, as detailed in **Figure 4.39**. Use a 5/32-inch bit to drill the holes. Once the pieces have been cut and drilled, bend the pieces, as shown by the arrows in **Figure 4.39**. Use a table vise or the edge of a table to bend the pieces. **Figure 4.40** shows a finished servo mount.

Z1 and Z2
Two pieces

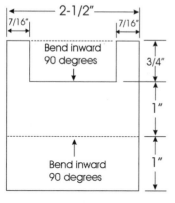

Cutting and bending guide

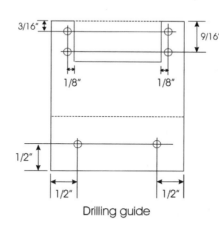

Drilling guide

FIGURE 4.39

Cutting, bending, and drilling guide for the servo mounts.

FIGURE 4.40

Finished servo mount.

To fabricate the two servo horn spring-loading mechanisms, take a piece of the 1/2-inch aluminum stock and cut two pieces to a length of 2 inches each. Cut and drill the pieces, as shown in **Figure 4.41**. The finished pieces should resemble the one shown in **Figure 4.42**. Modify two servo horns so that they resemble the one shown in **Figure 4.42**. This is accomplished by cutting two of the cross pieces off with a pair of side cutters, and then lining up the middle hole of the servo horn with the middle hole in piece W or X. When the middle holes are lined up, mark the area where the 5/32-inch holes line up. Use a 5/32-inch bit to drill on the markings so that the finished horn looks like the one in **Figure 4.42**. Attach one of the modified servo horns to piece W with two 6/32-inch × 1/2-inch machine screws and locking nuts. Attach the second servo horn to piece X, also using two 6/32-inch × 1/2-inch machine screws and locking nuts. Insert two 6/32-inch × 3/4-inch machine screws through the two outer holes of piece W, and secure in place with two lock washers and nuts. Do the same for piece X. One of the finished spring-loading mechanisms is shown in **Figure 4.43**, and can be used as a guide.

FIGURE 4.41

Cutting and drilling guide for spring-loading mechanism.

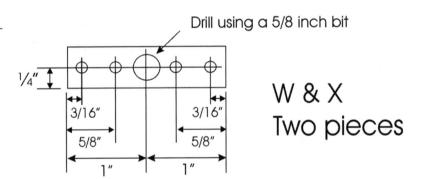

Drill using a 5/8 inch bit

1/4"

3/16"   3/16"

5/8"   5/8"

1"   1"

W & X
Two pieces

Drill other holes with a 5/32 inch bit

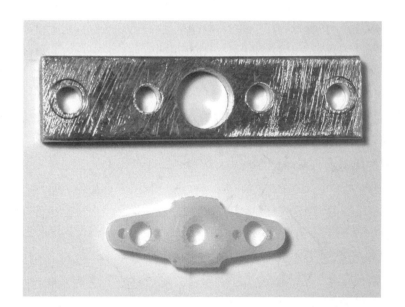

FIGURE 4.42

Spring-loading
mechanism and
modified servo horn.

FIGURE 4.43

Finished spring-loading
mechanism.

Take the two completed spring-loading mechanisms and attach each one to a servo that has been modified for continuous rotation. Use the servo screw that came with the servo horn to secure the mechanism in place on the servo shaft, as shown in **Figure 4.44**.

Take one of the servos with the spring-loading mechanism attached and secure it to a servo mount, using four 6/32-inch × 1/2-inch machine screws, lock washers, and nuts. Attach the second servo to the second servo mount, also using four 6/32-inch × 1/2-inch machine screws, lock washers, and nuts, but note that the servo is attached so that it mirrors the first one, as shown in **Figure 4.45**.

Attach the servo mounts to the frog's body using four 6/32-inch × 1/2-inch machine screws, lock washers, and nuts. The servo

FIGURE 4.44

Spring-loading mechanism attached to a modified servo.

FIGURE 4.45

Servos attached to servo mounts. Note the mirrored configuration.

mounts should be positioned with the servo shaft closest to the frog's head. **Figure 4.46** illustrates the proper orientation of the servo mounts on the frog's body.

FIGURE 4.46

Servo mounts attached to the frog's body.

Constructing the Front Legs

Cut two front leg pieces to a length of 4 inches, using the 1/2-inch aluminum. Use **Figure 4.47** as a guide to cut, drill, and bend the aluminum. Attach the finished legs to the robot's body, using two 6/32-inch × 1/2-inch machine screws, lock washers, and nuts, as shown in **Figure 4.48**.

FIGURE 4.47

Cutting, drilling, and bending guide for the front legs.

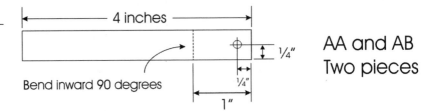

Holes are drilled with a 5/32 inch drill bit

FIGURE 4.48

Front legs attached to the robot's body.

Leg Position Sensors

The leg position sensors are limit switches that will determine when the legs are set to their jumping position, at which point the spring mechanism is fully loaded. This information will be used by the microcontroller to coordinate the legs for jumping. To attach the limit switches, manually rotate each servo by hand so that the spring is fully loaded toward the top of the spring-loading mechanism's travel, as shown in **Figure 4.49**. While maintaining this position, use hot glue to fix the limit switch to part E so that the switch is triggered, as shown in **Figure 4.49**. Do the same for the other leg, attaching the second limit switch to part F.

Wiring the Limit Switches

Cut a piece of 2-strand connector wire to a length of 6 inches and solder the wire to connect the two limit switches, as shown in **Figure 4.50**. Cut another piece of the 2-strand connector wire to a length of 3-1/2 inches. Solder one end of each wire to a 2-post female header connector, and the opposite ends to the left leg limit

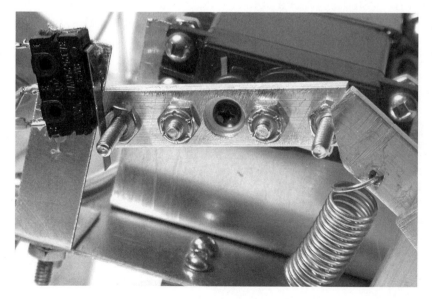

FIGURE 4.49

Limit switch hot glued to part E.

FIGURE 4.50

Limit switch wiring
diagram.

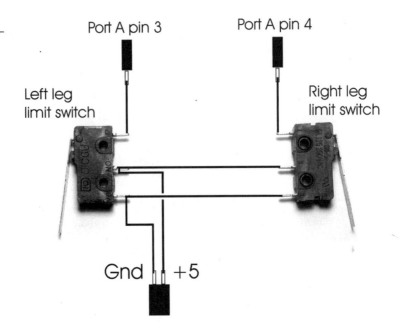

switch, as shown in **Figure 4.50**. The header will be plugged into
the +5 VDC and the GND connector on the main controller circuit
board later in the chapter.

Next, cut two single-strand connector wires to a length of 5-1/2
inches. Solder one end of each wire to a single-post female header connector, and the other end of each wire to the left and right
limit switches, as shown in **Figure 4.50**. The limit switch connectors will eventually be attached to microcontroller inputs. **Figure
4.51** shows the connectors wired to the limit switches.

Fabricate a 6-volt battery pack holder using 1/16-inch thick aluminum by following the cutting, drilling, and bending guide shown
in **Figure 4.52**. When the battery pack holder is finished, attach it
to the robot's body using a 6/32-inch × 1/2-inch machine screw,
lock washer, and nut. **Figure 4.53** shows the completed battery
pack holder fastened to the robot's body.

At this point, the robot's mechanical construction is complete. The
next section of Chapter 4 will focus on the electronics.

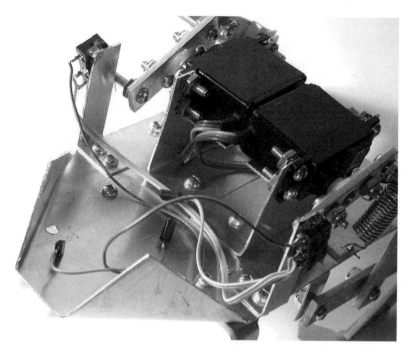

FIGURE 4.51

Limit switches wired to connectors.

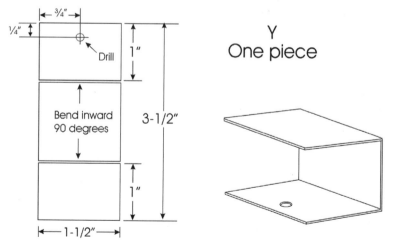

FIGURE 4.52

Cutting, drilling, and bending guide for the battery pack holder.

6 Volt battery pack holder

Finished 6 Volt battery pack holder

Hole is drilled with a 5/32 inch bit.

FIGURE 4.53

Battery pack holder fastened to the robot's body.

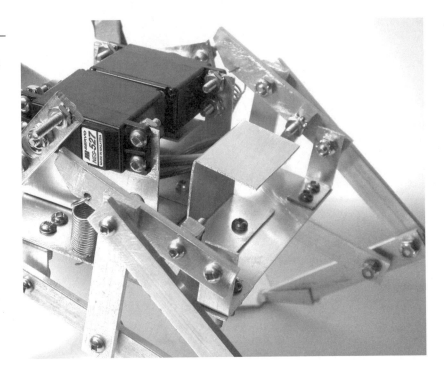

Frogbotic's Main Controller Board

This section focuses on the construction of the robot's main controller circuit and the fabrication of the printed circuit board (PCB). **Table 4.3** lists all of the parts necessary to build the controller board. All of the robot's functions are controlled by a Microchip PIC 16F84 microcontroller. The microcontroller is an entire computer on a chip, and makes it possible to eliminate a large amount of hardware that would otherwise be required. The microcontroller serves as the robot's "brain," controlling and managing all functions, sensors, and reflexes. The 16F84 microcontroller that we are using will be clocked at 4 MHz, and operates on a 5-volt DC supply, produced from a 78L05 voltage regulator, with the source being a 6-volt battery pack. The two leg servos are also powered by the same 6-volt DC battery pack. As you can see from the schematic shown in **Figure 4.54**, the input/output (I/O) lines are

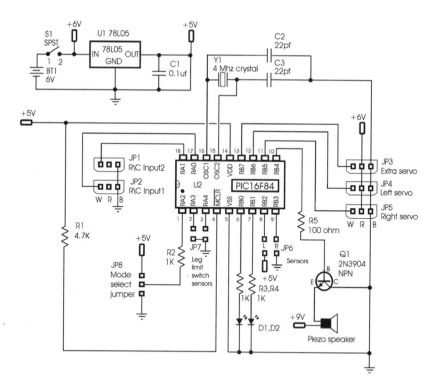

FIGURE 4.54

Frogbotic's main controller board schematic.

used as inputs and outputs to monitor the robot's leg position limit switches, turn on two light-emitting diodes (LEDs), and output sound to a piezo speaker. Each of the controller board's functions will be covered in detail when programming the robot.

Part	Quantity	Description
Semiconductors		
U1	1	78L05 5V regulator
U2	1	PIC 16F84 flash microcontroller mounted in socket
Q1	1	2N3904 NPN transistor
D1	1	Red light-emitting diode

TABLE 4.3

Parts List for Frogbotic's Main Controller Board

(continued on next page)

TABLE 4.3	Part	Quantity	Description
Parts List for Frogbotic's Main Controller Board (continued)	D2	1	Green light-emitting diode
	Resistors		
	R1	1	4.7 KΩ 1/4-watt resistor
	R2, R3, R4	3	1 KΩ 1/4-watt resistor
	R5	1	100 Ω 1/4-watt resistor
	Capacitors		
	C1	1	0.1 µf capacitor
	C2, C3	2	22 pf
	Miscellaneous		
	JP1–JP5, JP8	6	3-post header connector—2.5 mm spacing
	JP6, JP7	4	2-post header connector—2.5 mm spacing
	Battery	2	2-post header connector—2.5 mm spacing connectors
	Y1	1	4-MHz crystal
	Piezo buzzer	1	Standard piezoelectric element
	Battery holder	1	4-cell AA battery holder—6V output
	Battery strap	1	9V-type battery strap
	IC socket	1	18-pin IC socket—soldered to PC board U2
	Printed circuit board	1	See details in chapter.

Creating Frogbotic's Printed Circuit Board

To fabricate the PCB, photocopy the artwork in **Figure 4.55** onto a transparency. Make sure that the photocopy is the exact size of the original. For convenience, you can download the file from the

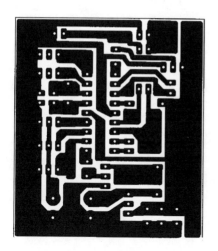

Figure 4.55

PCB foil pattern
artwork.

author's Web site, located at www.thinkbotics.com, and simply
print the file onto a transparency using a laser or ink-jet printer
with a minimum resolution of 600 dpi. After the artwork has been
successfully transferred to a transparency, use the techniques outlined in Chapter 2 to create a board. A 4-inch × 6-inch presensitized positive copper board is ideal. When you place the transparency on the copper board, it should be oriented exactly as in
Figure 4.55.

Circuit board drilling and parts placement. Use a 1/32-inch
drill bit to drill all of the component holes on the PCB. Drill the
holes for the voltage regulator (U1) with a 3/64-inch drill bit. Use
Table 4.3 and **Figure 4.56** to place the parts on the component
side of the circuit board. Note that the PIC 16F84 microcontroller
(U2) is mounted in an 18-pin I.C. socket. The 18-pin socket is soldered to the PC board and the PIC is inserted after it has been programmed. Use a fine-toothed saw to cut the board along the guide
lines and drill the mounting holes using a 6/32-inch drill bit.
Figure 4.57 shows the finished main controller board.

Check the finished board for any missed or cold soldered connections, and verify that all the components have been included. The

Figure 4.56

PCB component side
parts placement.

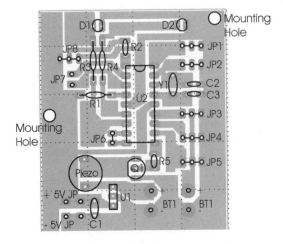

board will be tested later when programming the PIC microcontroller to coordinate the legs for jumping.

Fabricating the Power Connector

The next subassembly will be used to connect the battery pack to the controller board. **Table 4.4** lists the parts that will be needed.

Figure 4.57

Parts soldered to the
finished PCB.

Part	Quantity	Description	TABLE 4.4
Battery clip	1	Connects to the battery pack	List of Parts Needed to Fabricate the Power Connector
2-connector female header	2	2.5-mm spacing	
Switch	1	Single-pole single-throw 2-position toggle	
Connector wire	9 inches	18-gauge wire	
Battery pack	1	6V output 4-cell AA battery holder	
Nylon standoffs	2	1/4-inch diameter $\times$ 3/8-inch in length	
6/32 nylon machine screws	2	3/4-inch in length	
6/32 nylon nuts	2	Nylon nuts	

Solder the negative wire (black) of the battery clip to one of the terminals of the switch. Cut two pieces of connector wire to a length of 1 inch. Solder one end of each of the wires to each connector of a 2-terminal female connector. Solder the other end of each wire to each of the second 2-terminal female connectors. Cut a connector wire to a length of 7 inches, and solder one end to the other terminal of the switch. Solder the other end of the 7-inch wire to one of the connectors of one of the 2-terminal female connectors. Solder the positive (red) wire from the battery clip to the other terminal of the 2-terminal female connector. **Figure 4.58** shows how the battery clip, switch, and connectors are to be wired.

FIGURE 4.58

Finished power
connector.

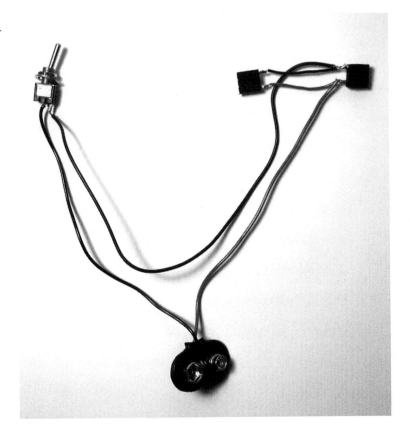

Putting It All Together

Fabricate two standoffs using 1/4-inch nylon or plastic tubing cut to a length of 3/8-inch. These will be used to raise the PCB up off of the robot's body when it is mounted. Place the standoffs between the mounting holes and the circuit board and secure in place with two 6/32-inch × 3/4-inch nylon machine screws and nuts. **Figure 4.60** shows the board mounted to the robot.

Follow the connection diagram in **Figure 4.59** to connect all of the individual components. The power connector cable that was just fabricated should be connected so that the female 2-post headers are plugged into the BT1 connectors, so that the terminals with the positive (red) battery lead are connected to the top posts. The switch is mounted in the 1/4-inch hole to the rear, right side of the body, and the battery clip should be positioned so that it is near the battery holder. **Figure 4.61** shows the power switch and the 6-volt battery pack hooked up to the battery clip. When the servos are plugged into the board, make sure that the yellow wires of the servo connectors are positioned to the inside of the board and the black wires are closest to the edge of the board. Connect the left and right limit switches to the controller board, as indicated in **Figure 4.59**. The completed frog robot is shown in **Figure 4.62**.

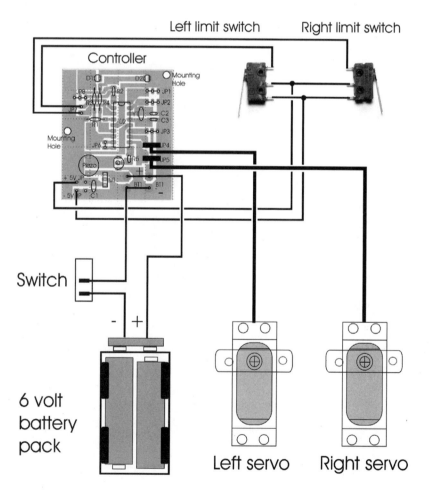

FIGURE 4.59

Frogbotic component connection diagram.

FIGURE 4.60

Controller board with
connectors attached.

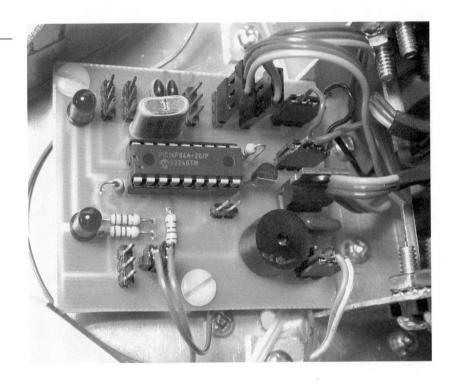

FIGURE 4.61

Power switch and 6-volt
battery pack.

FIGURE 4.62

Rear-side view of the finished robot frog.

Now that Frogbotic's hardware is complete, we will focus on programming the robot to read input from the limit switch sensors, control the leg servos, make sounds, and turn the LEDs on and off.

Programming and Experiments with Frogbotic

To test the main controller board, the PIC 16F84 will be programmed to flash the LEDs, make frog-like noises, and then start rotating the servos. This will ensure that all of the components have been correctly soldered to the board and that power has been connected. The first program is called frog-test.bas and is listed in **Program 4.1**. Type the program into your favorite text editor and then compile the code. Program the PIC 16F84, as detailed in Chapter 3, with the frog-test.hex file, listed in **Program 4.2**. When the chip has been successfully programmed, insert it into the 18-pin I.C. socket on the main controller board with the notch and pin 1 facing toward the LEDs and then apply power. If everything is working properly, the LEDs should flash on and off while making frog noises. When the light and sound stops, the servos should start rotating in a forward direction toward the front of the robot. If the servos are rotating in the opposite direction, then switch the two servo connectors on the controller board.

PROGRAM 4.1

frog-test.bas program listing

```
'_____
' Name       : Frog-test.bas
' Compiler : PicBasic Pro MicroEngineering Labs
' Notes      : Program to test the main controller
'            : board by flashing LEDs, producing
'            : sounds and slowly rotating the servos
'_____


' set porta to inputs
trisa = %11111111

' set portb pins 2 & 3 to inputs
trisb = %00001100


'_____

' initialize variables

servo_pos_l   VAR BYTE
servo_pos_r   VAR BYTE
timer1        VAR BYTE
timer2        VAR BYTE
timer3        VAR BYTE
temp1         VAR BYTE
servo_r       VAR PORTB.5
servo_l       VAR PORTB.6
switch_r      VAR PORTA.4
```

```
switch_l        VAR PORTA.3
led_l           VAR PORTB.1
led_r           VAR PORTB.0
piezo           VAR PORTB.4
```

PROGRAM 4.1

frog-test.bas program
listing (continued)

```
'_____


low servo_l
low servo_r

start:

for temp1 = 1 to 10
    SOUND piezo, [80,4,100,2]
    low led_l
    low led_r
    pause 50
    high led_l
    high led_r
next temp1

SOUND piezo, [100,4,120,2,80,2,90,2]

low led_l
low led_r

rotate:

servo_pos_r = 170
gosub right_servo

servo_pos_l = 130
gosub left_servo

goto rotate

'_____

' subroutines to set servos
```

PROGRAM 4.1

frog-test.bas program
listing (continued)

```
both_servo:
    for timer1 = 1 to 15
        pulsout servo_l,servo_pos_l
        pulsout servo_r,servo_pos_r
        pause 6
    next timer1
return

left_servo:
    for timer2 = 1 to 10
        pulsout servo_l,servo_pos_l
        pause 6
    next timer2
return

right_servo
    for timer3 = 1 to 10
        pulsout servo_r,servo_pos_r
        pause 6
    next timer3
return

end
```

PROGRAM 4.2

frog-test.hex program
listing

```
:100000007B28A0003B200C080D04031976287020E3
:100010008413200880066400 0D280E288C0A03191A
:100020008D0F0B28800676288F0022088400200977
:100030003C2084138F0803197628F03091000E08B5
:1000400080389000F0309103031 9910003198F0359
:10005000031976282B283F2003010C1820088E1F37
:1000600020088E0803190301900F382880061F28E6
:10007000392800002228FF3A8417800576280D08C9
:100080000C0403198C0A80300C1A8D060C198D068D
:100090008C188D060D0D8C0D8D0D76288F018E0020
:1000A000FF308E07031C8F07031C762803308D005A
:1000B000DF305C2050288D01E83E8C008D09FC303B
:1000C000031C65288C07031862288C0764008D0FB9
```

```
:1000D00062280C186B288C1C6F2800006F28080001
:1000E0008C098D098C0A03198D0A080083130313E8
:1000F0008312640008008316FF3085000C308600F0
:100100008312061383160613831286128316861231
:100110008312013 0A60064000B3026020318B028B9
:10012000630A2001030A00050308E0004301420A1
:1001300064308E0002301420861083168610831 2DD
:10014000061083160610323083124E208614831652
:10015000861083120614831606108312A60F8B28AE
:10016000630A2001030A00064308E00043014204D
:1001700078308E0002301420503 08E00023014206F
:100180005A308E0002301420861083168610831297
:10019000061083160610831 2AA30A50000218230B3
:1001A000A400ED20CC280130A70064001030270205
:1001B0000318EC2824088C008D01063084004030A0
:1001C000012025088C008D0106308400203001209C
:1001D00006304E20A70FD52808000130A800640083
:1001E000B3028020318FF2824088C008D010630EC
:1001F00084004030012006304E20A80FEF28080070
:10020000130A90064000B302902031812292508C7
:100210008C008D0106308400203001200 6304E20F5
:0A022000A90F02290800630013294A
:02400E00F53F7C
:00000001FF
```

The next experiment will be to read the limit switches, and then turn on the corresponding LED when a limit switch has been activated. Compile the limit-switch.bas program listed in **Program 4.3**, and then program the PIC 16F84 with the limit-switch.hex listed in **Program 4.4**. Insert the PIC into the 18-pin socket on the controller board and turn on the power. Use your finger to activate the left limit switch. When the switch is activated, the left LED should turn on. If the right LED turns on when the left switch is activated, then switch the pins that the limit switches are attached to on the controller board. Try the same procedure with the right limit switch. If the LEDs do not react when the switches are triggered, then go back and check the wiring.

PROGRAM 4.2

Frog-test.hex program listing (cotinued).

107

PROGRAM 4.3

limit-switch.bas program
listing

```
'_____

' Name    : Limit-switch.bas
' Compiler : PicBasic Pro MicroEngineering Labs
' Notes    : Program to monitor the status of the leg
'          : position limit switches and turn on the
'          : corresponding LED when triggered
'_____


' set porta to inputs
trisa = %11111111

' set portb pins 2 & 3 to inputs
trisb = %00001100


'_____

' initialize variables

servo_pos_l     VAR BYTE
servo_pos_r     VAR BYTE
timer1          VAR BYTE
timer2          VAR BYTE
timer3          VAR BYTE
temp1           VAR BYTE

servo_r         VAR PORTB.5
servo_l         VAR PORTB.6
switch_r        VAR PORTA.4
switch_l        VAR PORTA.3
led_l           VAR PORTB.1
led_r           VAR PORTB.0
piezo           VAR PORTB.4

low servo_l
low servo_r

'_____

start:
```

```
for temp1 = 1 to 5
SOUND piezo, [80,4,100,2]
low led_l
low led_r
pause 50
high led_l
high led_r
next temp1
SOUND piezo, [100,4,120,2,80,2,90,2]
low led_l
low led_r


right:


if switch_r = 1 then
   high led_r
else
   low led_r
endif


left:


if switch_l = 1 then
   high led_l
else
   low led_l
endif


goto right


end
```

PROGRAM 4.3

limit-switch.bas program listing (continued)

```
:1000000061288F0022088400200928208 4138F088B
:1000100003195C28F03091000E0880389000F03011
:1000200009103031991000 3198F0303195C28182801
:100030002B2003010C1820088E1F20088E0803199E
:100040000301900F252880060C28262800000F2881
```

PROGRAM 4.4

limit-switch.hex program listing

PROGRAM 4.4

limit-switch.hex program listing (continued)

```
:100050008417800055C280D080C0403198C0A803075
:100060000C1A8D060C198D068C188D060D08C0D35
:100070008D0D5C288F018E00FF308E07031C8F07CB
:10008000031C5C2803308D00DF3048203C288D01A4
:10009000E83E8C008D09FC30031C51288C070318A6
:1000A0004E288C0764008D0F4E280C1857288C1C86
:1000B0005B2800005B28080083130313831264008D
:1000C00008008316FF3085000C308600831206136B
:1000D0008316061383128612831686128312013044A
:1000E000A60064000630260203189628063 0A200F7
:1000F0001030A00050308E000430012064308E009B
:10010000023001208610831686108312061083 1693
:100110000610323083123A2086148316861083121A
:10012000061483160610831 2A60F71280630A2004B
:100130001030A00064308E000430012078308E0032
:10014000023001205030 8E00023001205A308E00E3
:10015000023001208610831686108312061083164 3
:100160000610831264000051EBA28061483160610B2
:100170008312BE2806108316061083126400851DA4
:10018000C6288614831686108312CA288610831602
:0A01900086108312B2286300CB280A
:02400E00F53F7C
:00000001FF
```

Now that everything is running correctly, it is time to put all of the individual pieces of software together into one program that will allow the frog robot to jump in a coordinated manner. The program will start by monitoring the limit switches that determine when the legs' spring-loading mechanisms are set in the proper position. If the limit switches are not triggered, then the program will command the servos to rotate forward until both legs are set. When both legs are in position, the servos are paused for a moment and then both are commanded to rotate forward at the same time. The spring mechanisms then let go, and the energy stored in the springs forces each leg down quickly, causing the frog's body to leap forward off the ground. Because it is impossible to get the two legs perfectly coordinated when the mechanism lets go, the robot does not always leap forward. This introduces an interesting random element, and this is the key to actually controlling the robot's in-flight direction. If you want to add sensors and direction control to the robot, try writing a routine that will let one of the legs release, and wait for a predetermined amount of time before letting the other leg go. These values could be dependent on the information that the microcontroller receives from the sensor input. Compile frogbotic.bas listed in **Program 4.5**, and then program the PIC 16F84 with the frogbotic.hex file listed in **Program 4.6**. Insert the PIC 16F84 into the 18-pin socket on the controller board. Place the robot on a flat surface and turn on the power. Note that the robot should not be used on hardwood or other types of flooring that scratches easily.

PROGRAM 4.5

frogbotic.bas program listing

```
'_____
' Name      : Frogbotic.bas
' Compiler  : PicBasic Pro MicroEngineering Labs
' Notes     : Program to coordinate the jumping of
'           : a robotic frog
'_____


' set porta to inputs
trisa = %11111111

' set portb pins 2 & 3 to inputs
trisb = %00001100


'_____

' initialize variables

servo_pos_l     VAR BYTE
servo_pos_r     VAR BYTE
timer1          VAR BYTE
timer2          VAR BYTE
timer3          VAR BYTE
temp1           VAR BYTE
```

PROGRAM 4.5

frogbotic.bas program
listing (continued)

```
servo_r      VAR PORTB.5
servo_l      VAR PORTB.6
switch_r     VAR PORTA.4
switch_l     VAR PORTA.3
led_l        VAR PORTB.1
led_r        VAR PORTB.0
piezo        VAR PORTB.4

low servo_l
low servo_r

'_____

start:

for temp1 = 1 to 5
   SOUND piezo, [80,4,100,2]
   low led_l
   low led_r
   pause 50
   high led_l
   high led_r
   next temp1
   SOUND piezo, [100,4,120,2,80,2,90,2]
   low led_l
   low led_r

right:

if switch_r = 1 then
   high led_r
   servo_pos_r = 158
   gosub right_servo
else
   low led_r
   servo_pos_r = 200
   gosub right_servo
```

PROGRAM 4.5

frogbotic.bas program
listing (continued)

```
endif

left:

if switch_l = 1 then
   high led_l
   servo_pos_l = 152
   gosub left_servo
else
   low led_l
   servo_pos_l = 100
   gosub left_servo
endif

if switch_l = 1 and switch_r = 1 then
   for temp1 = 1 to 6
   servo_pos_l = 150
   servo_pos_r = 159
   gosub both_servo
   next temp1
   servo_pos_l = 100
   servo_pos_r = 200
   gosub both_servo
endif

goto right


'_____

' subroutines to set servos

both_servo:
     for timer1 = 1 to 15
     pulsout servo_l,servo_pos_l
     pulsout servo_r,servo_pos_r
     pause 6
     next timer1
return
```

PROGRAM 4.5

frogbotic.bas program
listing (continued)

```
left_servo:
    for timer2 = 1 to 10
    pulsout servo_l,servo_pos_l
    pause 6
    next timer2
return

right_servo
    for timer3 = 1 to 10
    pulsout servo_r,servo_pos_r
    pause 6
    next timer3
return

end
```

PROGRAM 4.6

frogbotic.hex program
listing

```
:100000009728A4003B200C080D04031992288C208B
:100010008413240880066 4000D280E288C0A031916
:100020008D0F0B28800692288F0026088400240953
:100030003C2084138F0803199228F03091000E0899
:100040080389000F03091030319910003198F0359
:10005000031992282B283F2003010C1824088E1F17
:1000600024088E0803190301900F382880061F28E2
:100070003928000022 28FF3A8417800592280D08AD
:100080000C0403198C0A80300C1A8D060C198D068D
:100090008C188D060D0D8C0D8D0D92288F018E0004
:1000A000FF308E07031C8F07031C922803308D003E
:1000B000DF305C2050288D01E83E8C008D09FC303B
:1000C000031C65288C07031862288C0764008D0FB9
:1000D00062280C186B288C1C6F2800006F28080001
:1000E0008D018F018E000230752894000F080D02DB
:1000F000031D7C280E080C0204300318013003197C
:100100002301405031DFF3092280038031DFF3014
:100110000405031DFF3092288C098D098C0A0319F0
:100120008D0A080083130313831264000800831 6EA
:10013000FF3085000C308600831206138 3160613E9
:100140008312861283168612831 20130AA0064007D
```

```
:1001500006302A020318CC280630A6001030A4006E
:100160005030BE000430142064308E000230142091
:100170008610831686108312061083160610323OFE
:100180008312AE208614831686108312061483165B
:1001900006108312AAOFA7280630A6001030A4006C
:1001A00064308E000430142078308E000230142029
:1001B00050308E00023014205A308E00023014204D
:1001C0008610831686108312061083160610831278
:1001D0006400051EF32806148316061083129E3051
:1001E000A9006621FA280610831606108312C8306B
:1001F000A90066216400851D0529861483168610D2
:100200008312983OA80053210C298610831686107B
:100210008312643OA800532100308519013080000E
:1002200001307020 9E000030051A01308C00013032
:100230007020A0001E08840020088520A000A100D6
:1002400064002008210403193829013OAA00640041
:1002500007302A02031833299630A8009F30A900DE
:100260003921AA0F27296430A800C830A9003921F4
:10027000E8280130AB0064001030280203185229 2B
:10028000028088C008D010630840040300120 2908A8
:100290008C008D0106308400203001200 6304E2075
:1002A000AB0F3B2908000130AC0064000B302C027E
:1002B0000318652928088C008D0106308400403021
:1002C000012006304E20AC0F552908000130AD004A
:1002D00064000B302D020318782929088C008D0149
:1002E00006308400203001200 6304E20AD0F6829F2
:0602F000080063007929FB
:02400E00F53F7C
:00000001FF
```

All of the code above can be downloaded from the ThinkBotics Web site located at: www.thinkbotics.com. More Frogbotic experiments will be explored later in the book. Have fun experimenting with the frog, and don't be afraid to modify and improve the designs however you see fit. **Figure 4.63** shows Frogbotic leaping from one leg.

PROGRAM 4.6

frogbotic.hex program listing (continued)

FIGURE 4.63

Frogbotic leaping from
one leg.

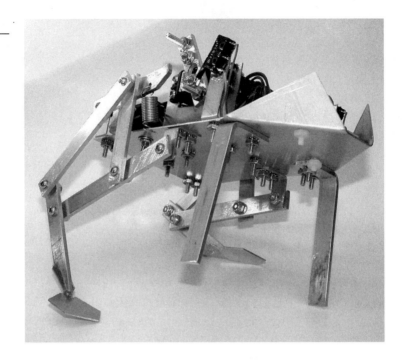

5

Serpentronic: Build Your Own Robotic Snake

Snakes

Snakes are characterized by a long, slender body covered in overlapping scales. There are approximately 2,800 species of snakes, most of which are nonvenomous and do not harm humans. They have no limbs or external ears. Snakes possess a backbone and ribs that may number in the hundreds. The eyelids do not move, but are fused to form transparent spectacles. The jaws of the mouth are not fused, which gives the snake the ability to open its mouth wide. This allows snakes to eat prey that are much bigger in diameter than they are. After the snake has swallowed, the bulge of the newly eaten animal can be seen before the snake's digestive process breaks it down. The snake's forked tongue is completely retractable. The snake's organs, such as the heart and stomach, are long and narrow. Only one lung is functional, with the left lung being unusable or missing entirely. Some primitive snakes have teeth only in one jaw, while the egg-eaters have no teeth at all.

Most snakes achieve locomotion by slithering along an S-shaped path. On land, a snake presses down and pushes forward from the curve of its body. The same slithering action also works well in the water. Sidewinders live much of their lives on sand. These snakes have developed a sideways movement because the sand slips away under them if they try to slither. A sidewinder throws a loop of its body forward. It then shifts its weight, raises its head and tail, and catches up to itself. Snakes move relatively slowly, and could not keep up with a person walking at a normal pace, which is about 4 miles per hour. The scales on a snake's body give them better traction as they slide along. They use rippling muscles in their bellies to shift their wide scales on edge. The edges catch on the ground and allow the snake to pull itself along.

The snake and its method of locomotion are the inspiration for the robot in this chapter. **Figure 5.1** shows a typical snake (Northern Death Adder), along with its biologically inspired robotic counterpart. The robot snake measures 28 inches in length, from head to tail, and is 2-1/2 inches wide. **Figure 5.2** illustrates the size of the snake relative to a human.

FIGURE 5.1

A snake and its biologically inspired robotic counterpart.

FIGURE 5.2

Robot snake showing size in relation to its human creator.

Overview of the Serpentronic Project

The robot snake that will be built and programmed in this chapter consists of six segments and a head, with each segment being powered by an R/C servo. The segments alternate in orientation so that the first segment moves in a horizontal motion and the next segment moves in a vertical motion. This sequence repeats itself for all six segments and the head, as shown in **Figures 5.3** and **5.4**. This gives the snake enough flexibility to move its body in a number of different ways in order to achieve locomotion, in much the same way as a biological snake. The robot is controlled by a Microchip PIC 16F84 microcontroller. The microcontroller is used to sequence the movement of each of the snake's body sections via servos. The microcontroller also monitors an infrared sensor so that the snake will avoid obstacles as it explores.

FIGURE 5.3

Diagram of the robot
snake's movement
capabilities.

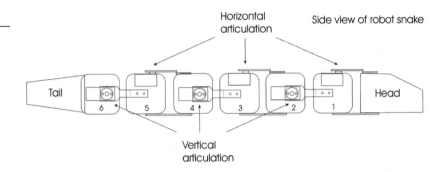

Horizontal
articulation

Side view of robot snake

Tail

Head

6 5 4 3 2 1

Vertical
articulation

FIGURE 5.4

Close-up view of the
snake's joints.

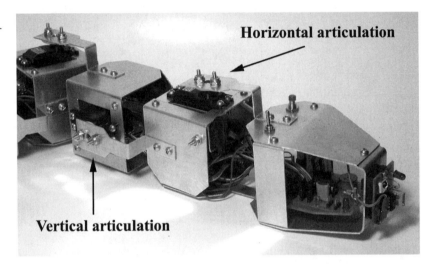

Horizontal articulation

Vertical articulation

Mechanical Construction of Serpentronic

The construction of the robot snake will begin with the mechanical construction of the body, head, and tail. The parts needed for
the mechanical construction are listed in **Table 5.1**.

TABLE 5.1

Parts List for
Serpentronic's
Mechanical
Construction

Parts	Quantity
1/16-inch thick aluminum stock	8-foot × 10-foot piece
6/32 × 1/2-inch machine screws	98
6/32 locking nuts	98
6/32 nylon washers	6
Standard R/C servo and hardware	6

The body, head, and tail are constructed using 1/16-inch flat aluminum.

Constructing the Body Sections

Start by cutting six pieces of the 1/16-inch aluminum to a size of 7-1/2 inches × 2-1/2 inches. These pieces will be identified as piece A of each of the six body sections. Use **Figure 5.5** as a guide to cut the six pieces to the dimensions shown. When the pieces are cut, use a 5/32-inch drill bit to drill the holes, as indicated in the diagram. File any rough edges from the pieces. Bend each

FIGURE 5.5

Cutting, bending, and drilling guide for body pieces (A).

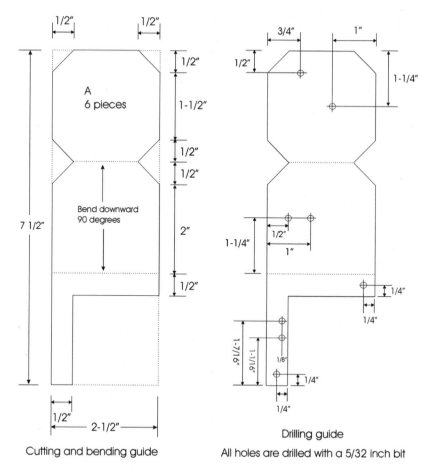

A
6 pieces

Bend downward
90 degrees

Cutting and bending guide

Drilling guide

All holes are drilled with a 5/32 inch bit

FIGURE 5.6

Finished body piece (A).

piece in a table vise or on the edge of a table, as indicated. Each of the six pieces should look like the one pictured in **Figure 5.6**.

The next piece that will make up each of the body sections is also cut from 1/16-inch thick aluminum. Cut six pieces to a size of 2-1/2 inches × 5-3/4 inches each. These pieces will be identified as piece B of each of the six body sections. Use **Figure 5.7** as a guide to cut the six pieces to the dimensions shown. When the pieces are cut, use a 5/32-inch drill bit to drill the holes, as indicated in the diagram. File any rough edges from the pieces. Bend each piece in a table vise or on the edge of a table, as indicated. Each of the six pieces should look like the one pictured in **Figure 5.8**.

B
6 pieces

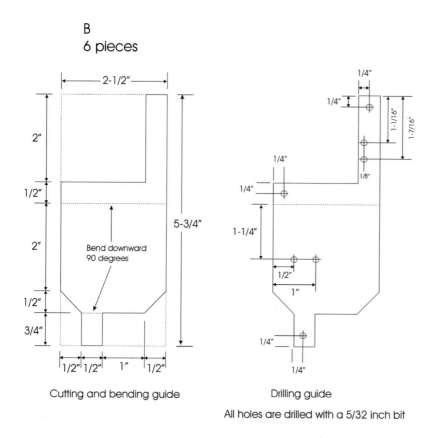

Cutting and bending guide

Drilling guide

All holes are drilled with a 5/32 inch bit

FIGURE 5.7

Cutting, bending, and drilling guide for body pieces (B).

FIGURE 5.8

Finished body piece (B).

Attach pieces A and B together using three 6/32-inch × 1/2-inch machine screws and locking nuts, as shown in **Figure 5.9**. Note that piece B is attached so that it is positioned on top of piece A. Continue the above procedure until all six body segments are completed.

FIGURE 5.9

Completed snake body segment made up of pieces A and B.

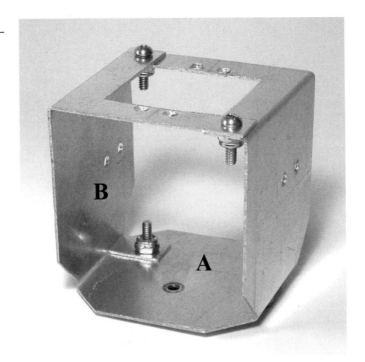

Attach a standard servo to each body segment using four 6/32-inch × 1/2-inch machine screws and locking nuts, as illustrated in **Figure 5.10**. When the servo is positioned in the body segment, the servo shaft side of the servo should be attached to piece B, as shown in **Figure 5.10**. Add standard servos to the remaining body segments using the procedure above.

FIGURE 5.10

Standard servo
attached to body
segment.

Cut six pieces of 1/16-inch thick aluminum to a size of 4-1/4 inches × 1 inch. Cut, drill, and bend each piece (C) according to **Figure 5.11**. The finished servo linkage pieces should resemble the one shown in **Figure 5.12**.

FIGURE 5.11

Cutting, bending, and drilling guide for servo linkage.

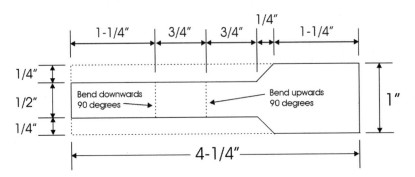

Cutting and bending guide

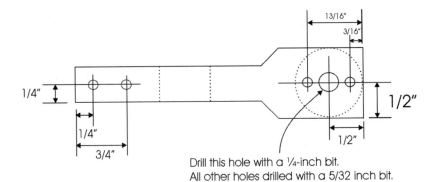

Drill this hole with a ¼-inch bit.
All other holes drilled with a 5/32 inch bit.

Drilling guide

FIGURE 5.12

Finished servo linkage.

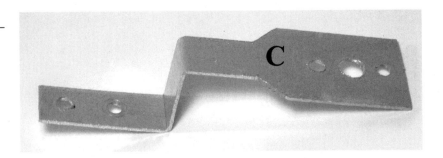

Take the circular, 1-inch diameter servo horn that came with your servo and line the middle mounting hole up with the 1/4-inch hole in piece C (see **Figure 5.11**). Mark the position on the servo horn where the two mounting holes line up, and drill them out with a 5/32-inch bit, as shown in **Figure 5.13**. Follow this procedure until a total of six servo horns are complete. Mount each of the completed servo horns to the six servo linkage pieces (marked C) using two 6/32-inch × 1/2-inch machine screws and locking nuts per linkage, as shown in **Figure 5.14**.

Cut six pieces of 1/16-inch thick aluminum to a size of 3-1/4 inches × 1/2-inch and drill as indicated in **Figure 5.15**. These six parts are identified as piece D, and are used as mechanical linkages to join each of the robot's body sections. Next, cut six pieces of 1/16-inch thick aluminum to a size of 1-1/2 inches × 1/2-inch, identified as piece E in **Figure 5.15**. Drill and bend each of the six

FIGURE 5.13

Servo horn with mounting holes.

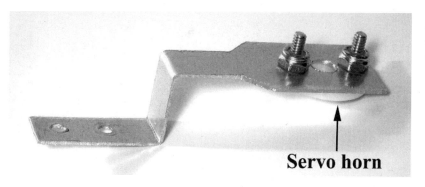

FIGURE 5.14

Servo horn attached to servo linkage.

Servo horn

E pieces, as shown in **Figure 5.15**. This part will be used to mount the battery holders in each of the body sections of the robot. **Figure 5.16** shows a completed mechanical linkage and battery pack mount.

FIGURE 5.15

Construction guide for mechanical linkage and battery pack mount.

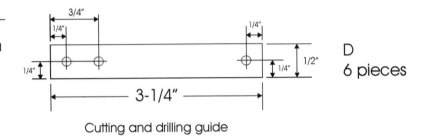

D
6 pieces

Cutting and drilling guide

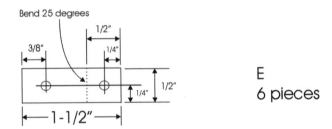

E
6 pieces

Cutting, bending and drilling guide
All holes are drilled with a 5/32 inch bit

FIGURE 5.16

Completed pieces D and E.

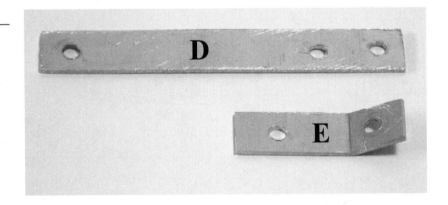

Take one of the 2-cell AA battery holders and drill a hole with a 5/32-inch bit, 5/8 of an inch from the edge of the holder without the battery clip connectors, as shown in **Figure 5.17**. Do this for all six of the 2-cell AA battery holders. Secure part E in place with a 6/32-inch × 1/2-inch machine screw and locking nut so that the bent part of piece E is to the left side of the battery pack, as shown in **Figure 5.18**. Do this for five of the holders. For the other holder that remains, secure part E in place so that the bent tab is oriented to the right. When this battery holder is connected to the tail section, it will be fastened differently than the rest.

Now that all of the individual mechanical pieces have been constructed, we will build the tail and head sections and then put it all together.

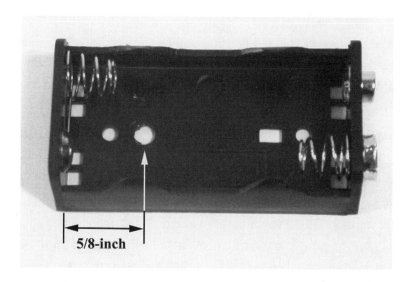

FIGURE 5.17

Battery pack drilling guide.

5/8-inch

FIGURE 5.18

Battery pack with
mounting piece
attached.

Constructing the Tail Section

The snake robot will need a tail that will be used to brace the rear
end of the body and provide friction when the robot is moving
forward and turning, as well as for the aesthetic purpose of com-
pleting the body.

The tail section is constructed using 1/16-inch thick aluminum
stock. Cut a piece 2-1/2 inches × 8-1/2 inches. File any rough
edges and place the piece on a table. Photocopy the cutting and
bending guide in **Figure 5.19**. Use the photocopier enlargement
feature so that the dimensions are exactly 2-1/2 inches × 8-1/2
inches. Cut the template out and use a glue stick to glue it onto the
piece of aluminum of the same size. Use a metal cutting band saw
or hacksaw to cut the piece, as shown in **Figure 5.19**. Drill the
mounting holes as indicated, using a 5/32-inch drill bit. Bend the
aluminum in a vise or on the edge of a table, as shown in **Figure
5.19**. The finished tailpiece is shown in **Figure 5.20**.

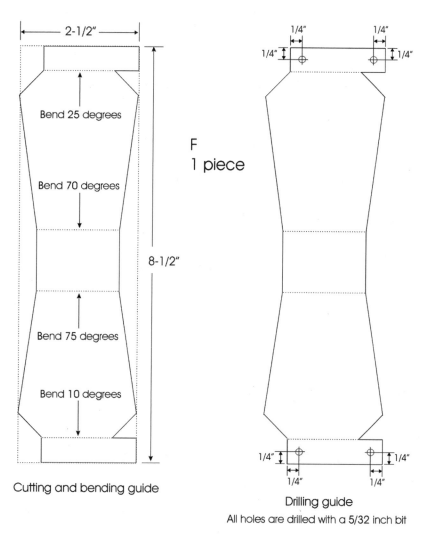

2-1/2"

Bend 25 degrees

Bend 70 degrees

8-1/2"

Bend 75 degrees

Bend 10 degrees

Cutting and bending guide

F
1 piece

1/4"   1/4"
1/4"   1/4"

1/4"   1/4"
1/4"   1/4"

Drilling guide
All holes are drilled with a 5/32 inch bit

FIGURE 5.19

Cutting, drilling, and bending guide for the snake's tail section.

FIGURE 5.20

Completed tail section.

Constructing the Head

The snake's head will house the controller board that will sequence all of the servos in each body section and will monitor the infrared sensor. The infrared sensor will also be mounted at the front of the head.

Cut a piece of 1/16-inch thick aluminum to a size of 3 inches × 6-1/4 inches. Cut, drill, and bend the piece, as shown in **Figure 5.21**. The finished piece, labeled G, is shown in **Figure 5.22**.

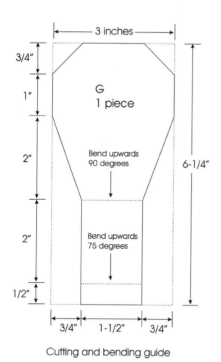

G
1 piece

Bend upwards
90 degrees

Bend upwards
75 degrees

3 inches

3/4"

1"

2"

2"

1/2"

6-1/4"

3/4" 1-1/2" 3/4"

Cutting and bending guide

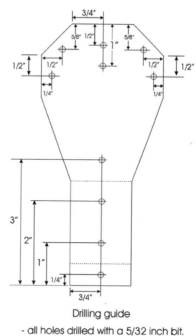

Drilling guide
- all holes drilled with a 5/32 inch bit.

FIGURE 5.21

Cutting, drilling, and bending guide for the bottom head piece G.

FIGURE 5.22

Finished head piece G.

Cut a piece of 1/16-inch thick aluminum to a size of 3 inches × 3-3/4 inches. Cut, drill, and bend the piece, as shown in **Figure 5.23**. The finished piece, labeled H, is shown in **Figure 5.24**.

FIGURE 5.23

Cutting, drilling, and bending guide for the top head piece H.

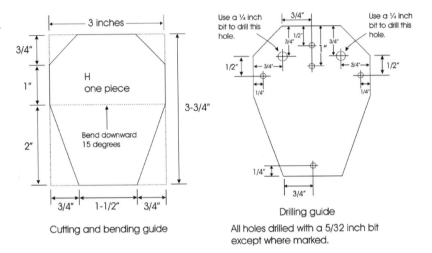

Cutting and bending guide

Drilling guide

All holes drilled with a 5/32 inch bit except where marked.

FIGURE 5.24

Finished top head piece H.

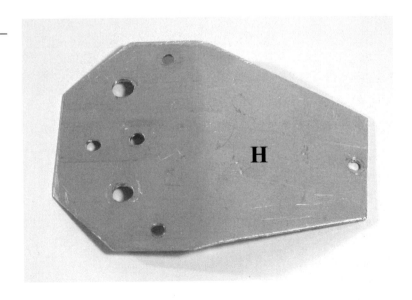

Cut two pieces of 1/16-inch aluminum to a size of 1 inch × 3-1/2 inches. Bend and drill each piece according to the dimensions shown in **Figure 5.25**. These two pieces are labeled I. The finished pieces are shown in **Figure 5.26** and will be used as the side supports for the robot's head.

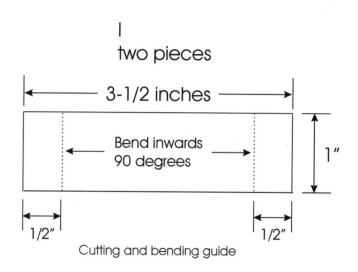

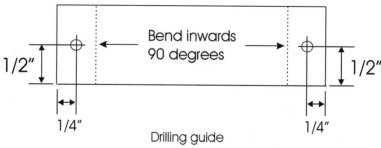

FIGURE 5.25

Cutting, drilling, and bending guide for head support pieces labeled I.

FIGURE 5.26

Finished head support pieces.

Each of the four head pieces will be assembled to form the robot's head. Use five 6/32-inch × 1/2-inch machine screws and locking nuts to assemble the head, as shown in **Figure 5.27**. Connect the two pieces labeled I to the bottom head piece labeled G. When those are secured, attach piece H to piece G, and the two pieces labeled as I.

FIGURE 5.27

Completed head assembly.

Assembling the Snake's Mechanical Structure

Now that all of the individual pieces that make up the snake's mechanical body have been constructed, it is time to put them all together.

Start by connecting the servo horn linkages made up of part C and a servo horn to each of the servos of each of the six body sections, as shown in **Figure 5.28**. Place the servo horn linkage onto the servo shaft without attaching the mounting screw. Turn the servo by hand all the way clockwise, and check to see if it is on a 90-degree angle from the center position. If it is not, then pull the servo horn linkage off and reattach it to the servo shaft at 90 degrees from the middle position. Turn the servo horn linkage all the way counterclockwise, and verify that it is also positioned on a 90-degree angle from the center position. Attach in place with the servo horn mounting screw that came with the servo. Follow this procedure for each of the six body sections.

FIGURE 5.28

Servo horn linkage attached to the servo.

Mount the mechanical linkage piece labeled D to each of the six body sections, as shown in **Figure 5.29**. This is accomplished by lining the single hole on the end of piece D up with the single hole on the body section (piece A) that is opposite to the servo. Secure in place with a 6/32-inch × 1/2-inch machine screw and locking nut with a 6/32-inch nylon washer between the mechanical linkage and the body section piece. The nylon washer acts as a bearing. Tighten the locking nut with enough torque to hold the parts in place, but allowing the piece to move freely. Repeat this same procedure for each of the six body sections.

FIGURE 5.29

Mechanical linkage attached to body section.

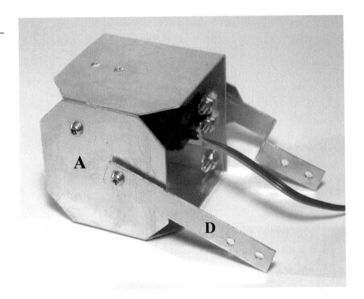

Connecting the Body Sections, Tail, and Head

At this point in the robot snake's construction, the serpent form starts to take shape. As each of the sections are joined, the battery packs will be added at the same time, since they will share the same fastener. Start with the section that will be the tail end of the snake. Locate the battery holder with the battery mounting connector attached to the opposite side, as all the others. Pick a body

FIGURE 5.30

Battery pack attached to body section 6.

section and connect the battery holder, as shown in **Figure 5.30**. Remove the locking nut that is connecting piece A and piece B of the body section. Connect the battery holder, and then secure in place with the locking nut that was just removed. This will be the body section that will have the tail section attached to it, and will be referred to as section 6.

Locate the tail section (piece F) and line it up to body section 6 so that the 1/2-inch section on either side overlaps on top of the body section by 1/2 an inch. Mark the location where the holes line up on the body section. Remove the tailpiece, and then drill the mounting holes marked on the body section with a 5/32-inch bit. Secure the tail piece in place with four 6/32-inch × 1/2-inch machine screws and locking nuts, as shown in **Figure 5.31**.

FIGURE 5.31

Tail connected to the
final body section.

Locate another one of the body sections and one of the battery
holders. Attach the mechanical linkage and the battery holder to
the body section using two 6/32-inch × 1/2-inch machine screws
and locking nuts, as shown in **Figure 5.32**. Next, attach the servo

FIGURE 5.32

Connected body
sections with battery
holder.

FIGURE 5.33

Alternating servo orientation of connected body sections.

linkage to the body section using two 6/32-inch × 1/2-inch machine screws and locking washers, as shown in **Figure 5.32**. Follow this same procedure for the rest of the body sections and battery holders. Note that each alternating body section will have the servo oriented to the snake's right side and then to the top, as illustrated in **Figure 5.33**.

The body segments alternate in orientation so that the first segment moves in a horizontal motion, and the next segment moves in a vertical motion. This sequence repeats itself for all six segments and the head. This gives the snake enough flexibility to move its body in a number of different ways in order to achieve locomotion, much the same way that a biological snake does.

Attach the head to body section 1 with four 6/32-inch × 1/2-inch machine screws and locking nuts, as shown in **Figure 5.34**. The head should be positioned so that the 1/4-inch mounting holes for the power switch and mode select push button are located on the top. Now that each of the body sections, head, and tail have been assembled, manually move each section through its range of motion to ensure that nothing obstructs the movement. Make any adjustments to the battery holders or mechanical linkages, if necessary.

FIGURE 5.34

Head section attached
to the robot's body.

Fabricate a 9-volt battery holder using 1/16-inch thick aluminum cut to a size of 4 inches × 1 inch. **Figure 5.35** is a cutting, drilling, and bending guide for the battery holder. When the battery holder is completed, attach it to the first body section behind the head. This is accomplished by positioning it in the top left corner of the body section and then marking the mounting hole. Drill out the hole in the body section with a 5/32-inch drill bit, and then mount the battery holder, as pictured in **Figure 5.36**. With this finished, the robot's mechanical construction is complete. Next, we will focus on fabricating the robot's main controller and infrared sensor circuit boards.

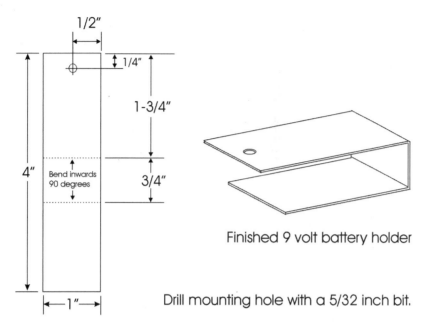

FIGURE 5.35

Cutting, bending, and drilling guide for 9-volt battery holder.

Finished 9 volt battery holder

Drill mounting hole with a 5/32 inch bit.

FIGURE 5.36

9-volt battery holder attached to the first body segment.

9 volt battery holder

Serpentronic's Main Controller Board

This section focuses on the construction of the robot's main controller circuit and the fabrication of the printed circuit board (PCB). **Table 5.2** lists all of the parts necessary to build the controller board. All of the robot's functions are controlled by a Microchip PIC 16F84 microcontroller. The microcontroller is an entire computer on a chip and makes it possible to eliminate a large amount of hardware that would otherwise be required. The microcontroller serves as the robot's "brain," controlling and managing all functions, sensors, and reflexes. The 16F84 microcontroller that we are using will be clocked at 4 MHz and operates on a 5-volt DC supply, produced from a 78L05 voltage regulator, with the source being a 9-volt battery. Each of the six servos used to move the body sections are powered by a separate 6-volt DC power source. The 6-volt power source is made up of the individual 3-volt battery packs in each of the body sections. As you can see from the schematic shown in **Figure 5.37**, the input/output (I/O) lines are

FIGURE 5.37

Serpentronic's main controller board schematic.

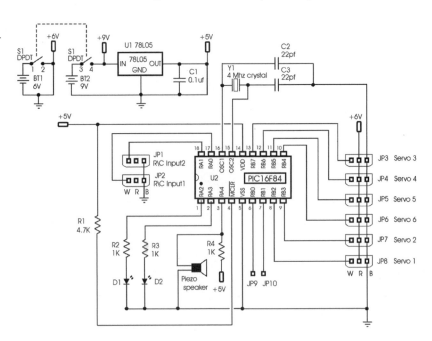

used to control the six servos, monitor the infrared sensor board, turn on two light-emitting diodes (LEDs), and output sound to a piezo speaker. Each of the controller board's functions will be covered in detail when programming the robot.

Part	Quantity	Description	
Semiconductors			**TABLE 5.2**
U1	1	78L05 5V regulator	Parts List for Serpentronic's Main Controller Board
U2	1	PIC 16F84 flash microcontroller mounted in socket	
D1	1	Red light-emitting diode	
D2	1	Green light-emitting diode	
Resistors			
R1	1	4.7 KΩ 1/4-watt resistor	
R2, R3, R4	3	1 KΩ 1/4-watt resistor	
Capacitors			
C1	1	0.1 μf capacitor	
C2, C3	2	22 pf	
Miscellaneous			
JP1–JP8	8	3-post header connector—2.5-mm spacing	
JP9, JP10	2	1-post header connector—2.5-mm spacing	
5-volt power	3	2-post header connector—2.5-mm spacing	
Y1	1	4-MHz crystal	
Piezo buzzer	1	Standard piezoelectric element	
BT1 and BT2	1	4-contact terminal block	
I.C. socket	1	18-pin I.C. socket—soldered to PC board U2	
Printed circuit board	1	See details in chapter.	

Creating the Main Controller Printed Circuit Board

To fabricate the printed circuit board (PCB), photocopy the artwork in **Figure 5.38** onto a transparency. Make sure that the photocopy is the exact size of the original. For convenience, you can download the file from the author's Web site, located at www.thinkbotics.com, and simply print the file onto a transparency using a laser or ink-jet printer with a minimum resolution of 600 dpi. After the artwork has been successfully transferred to a transparency, use the techniques outlined in Chapter 2 to create a board. A 4-inch × 6-inch presensitized positive copper board is ideal. When you place the transparency on the copper board, it should be oriented exactly as in **Figure 5.38**.

Figure 5.38

Controller board PCB foil pattern artwork.

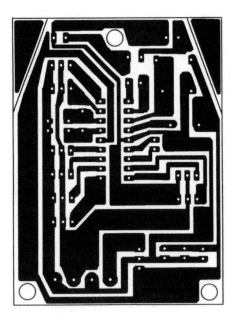

Circuit board drilling and parts placement. Use a 1/32-inch drill bit to drill all of the component holes on the PCB. Drill the holes for the voltage regulator (U1) with a 3/64-inch drill bit. Use **Table 5.2** and **Figure 5.39** to place the parts on the component side of the circuit board. Note that the PIC 16F84 microcontroller (U2) is mounted in an 18-pin I.C. socket. The 18-pin socket is soldered to the PC board and the PIC is inserted after it has been programmed. Use a fine-toothed saw to cut the board along the guide lines, and drill the mounting holes using a 5/32-inch drill bit. **Figure 5.40** shows the finished main controller board.

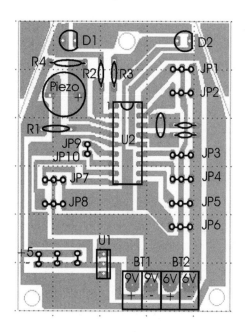

Figure 5.39

Controller board PCB component side parts placement.

Figure 5.40

Parts soldered to the
finished PCB.

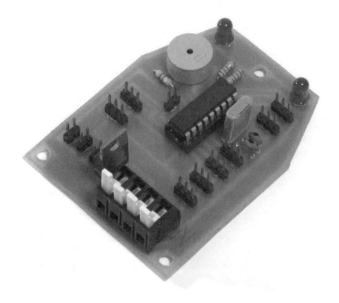

Check the finished board for any missed or cold soldered connections and verify that all the components have been included.
The board will be tested later when programming the PIC microcontroller.

The Infrared Sensor Board

An infrared sensor board will be fabricated to give the snake obstacle avoidance capabilities. The sensor board is comprised of an
infrared LED and a Panasonic PNA4602M IR sensor module. A sin-
gle-channel sensor is being used because the sensor board will be
mounted at the front of the robot's movable head. The snake is able
to move its head in an arc of 180 degrees, allowing it to sense objects
in front, and to either side of its body as it explores the surrounding
environment. The sensor board schematic is shown in **Figure 5.41**.
Table 5.3 is a list of all the parts needed to construct the board.

The 555 timer in the circuit is used to modulate the infrared LED
at a frequency determined by C1 and R3. R3 is an adjustable 10k

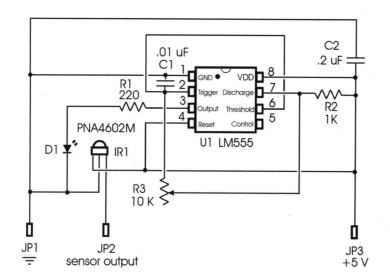

FIGURE 5.41

Infrared sensor board schematic.

potentiometer that will be used to find the optimum frequency during calibration. In our application, we will use a frequency between 38 and 42 kHz, so that a meaningful signal will be sent from the PNA4602 sensor module to the microprocessor.

The PNA4602M shown in **Figure 5.42** is designed to detect only infrared radiation that is modulated at 38 kHz, and rejects all other light sources. This makes the module an ideal sensor for daylight conditions. The features include an extension distance of 8 meters or more. No external parts are required, and a resin filter makes the module unsusceptible to visible light. **Table 5.4** lists the PNA4602M module's main characteristics. The output signals from the module will be processed and filtered by the microcontroller with a software routine, described later in the chapter.

TABLE 5.3	Part	Quantity	Description
Parts List for the Infrared Sensor Board	**Semiconductors**		
	U1	1	LM 555 Timer integrated circuit
	IR1	1	Panasonic PNA4602M infrared detector modules
	D1	1	Infrared light-emitting diodes
	Resistors		
	R1	1	220 Ω 1/4-watt resistor
	R2	1	1 KΩ 1/4-watt resistor
	R3	1	10 KΩ-ohm adjustable potentiometer
	Capacitors		
	C1	1	.01 µfd capacitor
	C2	1	.2 µfd capacitor
	Miscellaneous		
	JP1, JP2, and JP3	1	3-post header connector
	Printed circuit board	1	See details in chapter.
	I.C. socket	1	8-pin I.C. socket soldered to PC board to mount U1

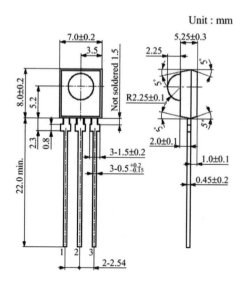

Unit : mm

1: V$_{OUT}$
2: GND
3: V$_{CC}$

FIGURE 5.42

Diagram of PNA4602M infrared sensor module.

Parameter	Symbol	Minimum	Typical	Maximum	Unit
Operating supply voltage	Vcc	4.7	5.0	5.3	V
Current consumption	Icc	1.8	2.4	3.0	mA
Max. reception distance	Lmax	8	10		m
Low-level output voltage	Vol		0.35	0.5	V
High-level output voltage	Voh	4.8	5.0	Vcc	V
Low-level pulse width	Twl	200	400	600	µs
High-level pulse width	Twh	200	400	600	µs
Carrier frequency	Fo		38.0		kHz

TABLE 5.4

Characteristics of the PNA4602M Module

The sensor board works by producing modulated infrared radiation with an infrared LED and using the PNA4602 module to detect any radiation reflected from the surface of solid objects. The PNA4602 sensor is designed to respond only to infrared that is modulated at a frequency somewhere between 38–42 kHz. The circuit is tuned to modulate the infrared LED at this frequency. Depending on the proximity of the sensor to the object, a greater or lesser number of infrared pulses will be reflected back. The number of reflected "hits" that the sensor receives in a given time frame allows the robot to determine how close it is to objects. The higher the number of reflected pulses, the closer the sensor is to the object. The output pin from the PNA4602 is connected to a microcontroller input pin, and a software routine is used to monitor the sensor.

Constructing the Infrared Sensor Circuit Board

To fabricate the PCB, photocopy the artwork in Figure 5.43 onto a transparency. Make sure that the photocopy is the exact size of the original. After the artwork has been successfully transferred to a transparency, use the techniques outlined in Chapter 2 to create a board. A 4-inch × 6-inch presensitized positive copper board is ideal. When you place the transparency on the copper board, it should be oriented so that it is exactly the same as in **Figure 5.43**.

FIGURE 5.43

Infrared sensor board PCB foil pattern artwork.

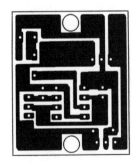

Circuit board drilling and parts placement. Use a 1/32-inch drill bit to drill all of the component holes on the PCB. Drill the holes for the voltage regulator (U1) with a 3/64-inch drill bit. Use **Table 5.3** and **Figure 5.44** to place the parts on the component side of the circuit board. Note that the 555 timer is mounted in an 8-pin I.C. socket. The 8-pin socket is soldered to the PC board and the 555 is inserted after the board has been soldered. Use a fine-toothed saw to cut the board along the guide lines, and drill the mounting holes using a 6/32-inch drill bit. **Figure 5.45** shows the finished main controller board.

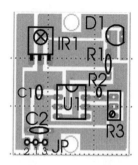

FIGURE 5.44

Infrared sensor board PCB component side parts placement.

FIGURE 5.45

Parts soldered to the finished PCB.

Calibration

To calibrate the infrared sensor board, a multimeter with frequency measuring capabilities like the one shown in **Figure 5.46** will be used. Connect a 5-volt DC source to the circuit, as shown in **Figure 5.47**. Connect the positive lead of the multimeter to the point shown in **Figure 5.47**, and connect the common lead to the circuit ground. Set the multimeter to read frequency. Use a small screwdriver to adjust potentiometer R3, until a frequency of approximately 43 kHz is displayed. This will adjust the circuit so that the 555 timer is producing a 5-volt square wave that will modulate the infrared LEDs at a frequency that is close to where the PNA4602 sensor module will respond. The circuit frequency will be fine-tuned with a software routine later in the chapter.

FIGURE 5.46

Fluke 87 digital multimeter with frequency measuring capabilities.

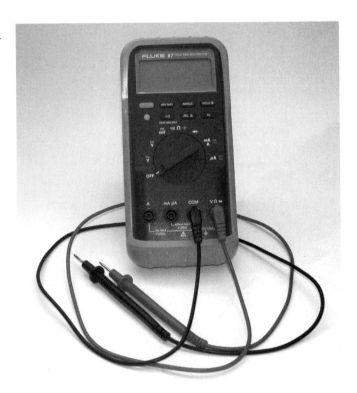

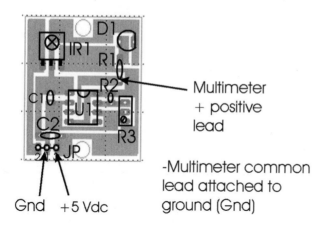

Multimeter + positive lead

-Multimeter common lead attached to ground (Gnd)

Gnd   +5 Vdc

FIGURE 5.47

Multimeter probe connection guide.

Mounting the Controller and Infrared Sensor Board

The main controller circuit board will be mounted in the snake's head on three 1/4-inch diameter nylon standoffs cut to a length of 1/2-inch. Position the standoffs over the mounting holes in the head and place the circuit board on top of the standoffs. Secure the board in place with three 6/32-inch × 1-inch machine screws, lock washers, and nuts, as shown in **Figure 5.48**. **Table 5.5** lists all the parts needed to mount the boards and wire the infrared sensor to the main controller board.

Part	Quantity	Description
3-strand ribbon wire	1	6 inches
3-connector female header	1	2.5-mm spacing
2-connector female header	1	2.5-mm spacing
1-connector female header	1	2.5-mm spacing
Heat shrink tubing	1	3 inches
1/4-inch diameter nylon standoff	5	1/4-inch length

TABLE 5.5

List of Parts Needed to Mount the Circuit Boards

(continued on next page)

TABLE 5.5

List of Parts Needed to
Mount the Circuit
Boards (continued)

Part	Quantity	Description
6/32 × 1-inch machine screws	3	
6/32 × 1-inch nut	3	
6/32 × 1-inch lock washer	3	
6/32 × 1/2-inch machine screw	2	
6/32 × 1/2-inch locking nut	2	

FIGURE 5.48

Controller circuit board
mounted in the snake's
head.

Cut a piece of 3-strand ribbon wire to a length of 6 inches. Strip
the ends and place a 1/2-inch length of heat-shrink tubing over
each wire. Solder the wires at one end to a 3-connector female
header and then shrink the tubing in place over the solder connections. On the other end of the wire, solder a single-connector
female header to one of the outside wires. Solder a 2-connector

FIGURE 5.49

Infrared sensor
connector wire.

female header to the other two wires. Use a heat source to shrink
the tubing over the solder connections. The finished connector
wire should resemble the one shown in **Figure 5.49**.

Mount the infrared sensor board to the front of the snake's head
on two 1/4-inch diameter nylon standoffs cut to a length of 1/4-
inch. Use two 6/32-inch × 1/2-inch machine screws and locking
nuts to secure the board in place, as shown in **Figure 5.50**. **Figure
5.51** is a wiring diagram showing how the connection wire should
be attached.

FIGURE 5.50

Infrared sensor board
attached to the front of
the snake's head.

FIGURE 5.51

Infrared sensor board
connection diagram.

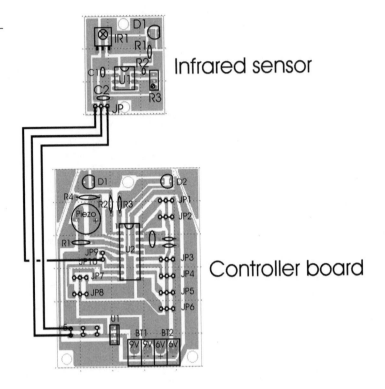

Infrared sensor

Controller board

Wiring the Robot

Next, the 3-volt battery packs, located in each body segment will
be wired to provide 6 volts to the controller board. The 6-volt supply will be used to directly power the servos. To accomplish this,
the first two battery packs will be wired in a series to create 6
volts. The next pair of battery packs are also wired in a series to
create 6 volts, as are the last two. Each of these three pairs are
then wired in parallel so that the supply is 6 volts, but capable of
providing higher current and a longer robot operating time. This is
important since the robot will be coordinating the movement of six
servos that may all be in operation at the same time. The 9-volt
supply is from a single battery mounted in the first body segment.
This supply is used to power the controller board. The use of dual
power supplies with a robot is preferred because it provides the

microprocessor with isolation from the noise introduced by the direct current motors in the servos. It also allows the robot to run for a much longer time because the microcontroller can keep operating from the 9-volt supply, even if the 6-volt supply drops down to 4 volts. The servos are capable of operating at lower voltages, but if the PIC's supply drops below 5 volts, it will go into a resetting loop. By powering the microcontroller with its own 9-volt source, this problem is eliminated.

Table 5.6 is a list of all the parts needed to complete the wiring of the robot snake.

Part	Quantity	Description	
Connector wire	1	3 feet	**TABLE 5.6** List of Parts Required to Wire the Robot
Battery straps, 9-volt type	7	Battery straps with 8-inch leads	
DPDT switch	1	Double-pole double-throw switch	
Push button switch	1	Momentary contact switch	
12-inch servo connector extension	3	Male and female connectors	
2-connector female header	1	2.5-mm spacing	
1-connector female header	1	2.5-mm spacing	
1-KΩ resistor	1	1/4-watt	
Rubber grippers	14	Sticky backed nonslip rubber	
AA battery	12	1.5-volt battery	
9-volt battery	1	9-volt battery	

Refer to **Figure 5.52** when wiring each of the 3-volt battery packs and the 9-volt battery to the DPDT switch and the controller board. Start by mounting the DPDT switch in one of the 1/4-inch mounting holes on the top of the snake's head. Wire each of the 3-volt battery packs in the body sections with the battery clips that attach to each holder. It may be easiest to connect each of the battery clips together before attaching them to the battery packs. Make sure that the connections are soldered in place and that insulating heat-shrink tubing is placed around each connection. All of the wires should run inside the snake from one section to another. Use the connector wire to attach the switch to the power terminal blocks on the controller board. Place a 9-volt battery in the battery holder that is located in the first body section behind the head (see **Figure 5.36**). Attach the negative lead of the 9-volt battery clip to the 9-volt power terminal connector on the controller board, and solder the positive lead to the switch.

FIGURE 5.52

Electrical wiring diagram for Serpentronic.

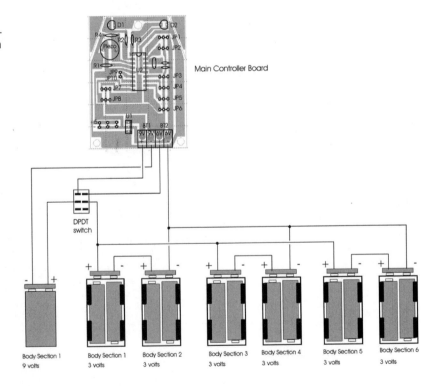

Next, connect all of the servos to the robot controller board located in the head. **Figure** 5.53 shows a diagram of the snake and each servo located in each body section, along with the corresponding connector on the controller board. When attaching each servo to the connector on the controller board, the servo's yellow wire is always to the left, as indicated in **Figure** 5.53. The middle wire is red and the wire to the right side is black. The last three servos labeled servo 4, servo 5, and servo 6 need wire extension connectors added so that they are long enough to reach the controller board. Use 2-inch servo connector extension wires, like the one shown in **Figure** 5.54.

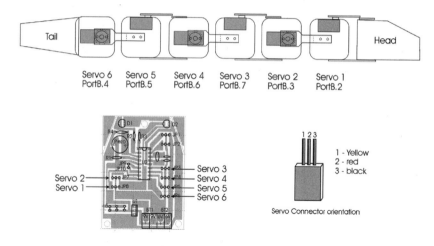

FIGURE 5.53

Servo connection diagram.

FIGURE 5.54

Servo connector extension wire.

A mode select push button will be added to the top of the robot's head so that different functions can be chosen when the robot starts up. One of the main functions that the push button will enable is calibrating the infrared sensors.

Fabricate the push button connector using a momentary contact switch, a 1-KΩ resistor, a 2-connector female header, a 1-connector female header, some heat-shrink tubing, and three pieces of connector wire cut to a length of 3 inches each. Use **Figure 5.55** as a wiring and soldering guide when creating the connector. When the push button assembly is finished, mount the push button in the 1/4-inch mounting hole on the snake's head and attach the connectors to the controller board, as shown in **Figure 5.56**.

FIGURE 5.55

Mode select switch wiring guide and finished assembly.

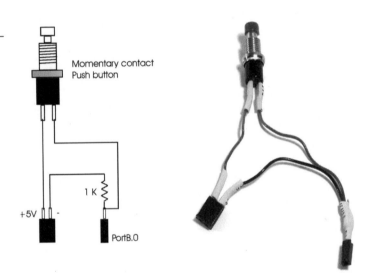

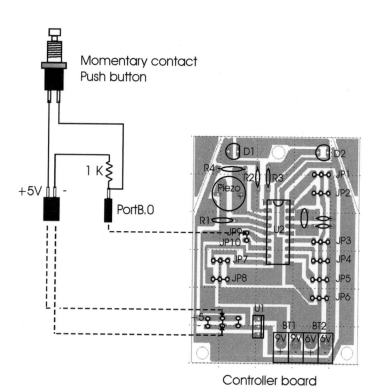

Momentary contact
Push button

1 K

+5V

PortB.0

Controller board

FIGURE 5.56

Mode select controller
board connection
diagram.

To give the snake added friction when moving, rubber gripper
pieces can be added to the underside of the snake's body. If you
decide not to add the rubber pieces, the robot will still function
properly, but will not move as easily on slippery surfaces like car-
pet. Any sort of rubber nonslip pieces that you can find at a hardware store will be suitable. The ones that I used have a sticky back
and were meant for the bottom of Sun System computers to stop
them from slipping on a desktop. **Figure 5.57** shows the positions
that worked well for me. Make sure that the movement of each
body segment is not hindered by the rubber gripper pieces.

FIGURE 5.57

Rubber gripper pieces added to the underside of the snake.

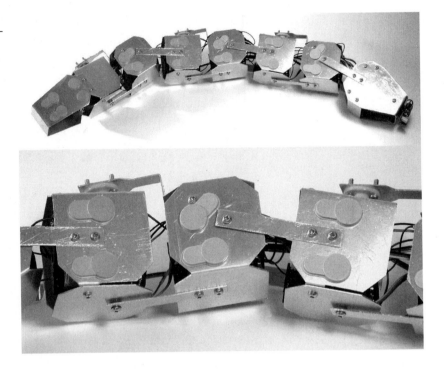

Insert two 1.5-volt, AA type batteries into each of the battery holders in each body segment. Add a 9-volt battery to the battery holder in the first body segment behind the robot's head and attach the battery strap. The robot is now ready to test and calibrate!

Programming and Experiments with Serpentronic

To test the main controller board, the PIC 16F84 will be programmed to flash the LEDs, make some random sounds, and center all of the servos. This will ensure that all of the components have been correctly soldered to the circuit board, and that the servos and batteries are all connected properly. The first program is called snake-test.bas and is listed in **Program 5.1**. Type the program into your favorite text editor, save, and then compile the program with PicBasic Pro, using the instructions in Chapter 3.

Program the PIC 16F84 with the snake-test.hex file listed in
Program 5.2. When the chip has been successfully programmed,
insert it into the 18-pin I.C. socket on the main controller board
with the notch and pin 1 facing toward the LEDs and then turn the
power switch to the "ON" position. If everything is working properly, the LEDs should flash on and off while making random noises. When the light and sound stops, the servos should all move to
the middle position, making the snake straight. If the snake is not
relatively straight, keep the power turned on and readjust each of
the servo horns so that it is straight. If nothing happens when
power is applied, then check all of the battery and circuit board
connections. Also, make sure that the PIC 16F84 was programmed
properly.

```
'_____
' Name      : Snake-test.bas
' Compiler  : PicBasic Pro - MicroEngineering Labs
' Notes     : Program to test the main controller
'           : board by flashing the LEDs, producing
'           : sounds and setting each of the servos to
'           : their middle positions
'_____


' PortA set as outputs
trisa = %00000000

' PortB set as outputs. pins 0-1 inputs
trisb = %00000011

'_____
' initialize variables

led_left    VAR PORTA.2
led_right   VAR PORTA.3
piezo       VAR PORTA.4
servo_1     VAR PORTB.2
```

PROGRAM 5.1

snake-test.bas program
listing

PROGRAM 5.1

snake-test.bas program
listing (continued)

```
servo_2      VAR PORTB.3
servo_3      VAR PORTB.7
servo_4      VAR PORTB.6
servo_5      VAR PORTB.5
servo_6      VAR PORTB.4
rand         VAR WORD
timer        VAR BYTE
temp1        VAR BYTE
i            VAR BYTE
servo1       VAR BYTE
servo2       VAR BYTE
servo3       VAR BYTE
servo4       VAR BYTE
servo5       VAR BYTE
servo6       VAR BYTE

low led_left
low led_right
Low servo1
Low servo2
Low servo3
Low servo4
Low servo5
Low servo6

'_____

' create randon noises and flash LEDs

    For temp1 = 1 to 7
      High led_left
      Low led_right
      GoSub randomize
      Pause 50

      Low led_left
      High led_right
      GoSub randomize
      Pause 50
```

```
    Next temp1

    Low led_right
```

PROGRAM 5.1

snake-test.bas program
listing (continued)

```
'_____

' start main execution

start:

    servo1 = 150
    servo2 = 150
    servo3 = 150
    servo4 = 150
    servo5 = 150
    servo6 = 150

    GoSub servo

goto start

'Subroutines start here


'_____

' random sound generator subroutine

randomize:

Random rand
i = rand & 31 + 64
Sound piezo,[i,4]
Return


'_____

' subroutine to set servos

servo:

    For timer = 1 to 20
```

167

PROGRAM 5.1

snake-test.bas program
listing (continued)

```
PulsOut servo_1,servo1
PulsOut servo_2,servo2
PulsOut servo_3,servo3
PulsOut servo_4,servo4
PulsOut servo_5,servo5
PulsOut servo_6,servo6
Pause 12
Next timer

Return
```

PROGRAM 5.2

snake-test.hex file
listing

```
:100000009128A0003F200C080D0403198C2886209D
:1000100084132008800664000D280E288C0A03191A
:100020008D0F0B2880068C288F0022088400200961
:100030000402084138F0803198C28F03091000E089B
:1000400080389000F03091030319910003198F0359
:1000500003198C282B28552003010C1820088E1F0B
:1000600020088E0803190301900F382880061F28E6
:100070003928000022284320FF3A80054028FF3A13
:1000800084178005 8C2894000063094190530840 06C
:1000900000308A00140807398207013402340 4341E
:1000A000083410342034403480340D080C04031913
:1000B0008C0A80300C1A8D060C198D068C188D0652
:1000C0000D0D8C0D8D0D8C288F018E00FF308E074D
:1000D000031C8F07031C8C2803308D00DF30722037
:1000E00066288D01E83E8C008D09FC30031C7B28BE
:1000F0008C07031878288C0764008D0F78280C185B
:1001000081288C1C85280000852808008C098D0911
:100110008C0A03198D0A08008313031383126400E9
:1001200008008316850103308600831205118316AB
:100130000511831285118316851183122708 3B2030
:1001400028083B2029083B202A083B202B083B207D
:100150002C083B200130AD0064000 8302D0203184C
:10016000C9280515831605118312851183168511 7B
:100170008312DB20323064200511831605118312AF
:100180008515831685118312DB2032306420AD0F74
:10019000AC2885118316851183129630A7009630FE
```

```
:1001A000A8009630A9009630AA009630AB00963091
:1001B000AC00F020CD2824088C0025088D005520A7
:1001C0000C08A4000D08A5005F302405A60005302A
:1001D000A2001030A00026088E0004301420080071
:1001E0000130AE00640015302E02031825292708BF
:1001F0008C008D0106308400043001202808C0001A
:100200008D0106308400083001202908C008D0102
:100210000630840080300120 2A088C008D010630D1
:10022000840040300120 2B088C008D0106308400B2
:100230000203001202C088C008D0106308400103005
:0C02400001200C306420AE0FF2280800F2
:02400E00F53F7C
:00000001FF
```

PROGRAM 5.2

snake-test.hex file
listing (continued)

The next program will be used to calibrate the infrared sensor so that the robot can safely avoid obstacles. The modulation frequency of the sensor was set using a multimeter when the circuit board was initially built. The software calibration routine will be used to fine-tune the frequency to improve the sensor's response. The routine works by taking the input from the sensor and then outputting the opposite state to the LEDs. The sensor input value is inverted before being output to the LEDs because the sensor's output is normally logic 1 (high) when it is not receiving a signal, and switches to a state of logic 0 (low) when a signal is received. This will allow us to visually see how the sensor is responding to the modulated infrared radiation, and then adjust the modulation frequency accordingly. The program is called ircal-serpent.bas, and is listed in **Program 5.3**. Program the PIC 16F84 with the corresponding ircal-serpent.hex file listed in **Program 5.4** and insert it into the 18-pin socket on the controller board. When the power is turned on and nothing is in front of the sensor, the LEDs should be off. To calibrate the circuit, use a small screwdriver to turn potentiometer R3 counterclockwise until the LEDs are on solid. **Figure 5.58** shows resistor R3 on the infrared circuit board being adjusted. Once the LEDs are on solid, slowly rotate potentiometer R3 clockwise until

FIGURE 5.58

Adjusting the infrared
sensor board.

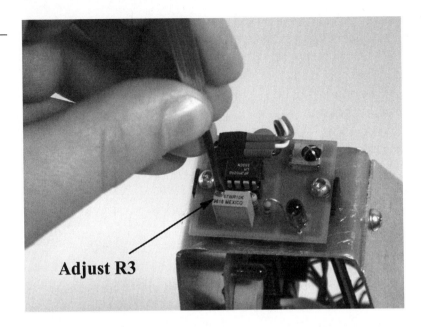

Adjust R3

the LEDs flicker and then turn off. Move your hand toward the sensor. With your hand at a distance of approximately 7 inches from the sensor, the LEDs should start to flicker. With your hand at a distance between 5 and 6 inches, the LEDs should be turned on solid.

PROGRAM 5.3

ircal-serpent.bas
program listing

```
'_____

' Name    : ircal-serpent.bas
' Compiler : PicBasic Pro - MicroEngineering Labs
' Notes    : Infrared sensor calibration program
'_____


' PortA set as outputs
trisa = %00000000

' PortB set as outputs. pins 0-1 inputs
trisb = %00000011


'_____

' initialize variables
```

PROGRAM 5.3

ircal-serpent.bas
program listing
(continued)

```
led_left        VAR portA.2
led_right       VAR portA.3
ir_input        VAR portB.1


low led_left
low led_right

ir_cal:

    If ir_input = 0 then
        high led_left
        high led_right
    endif
        low led_left
        low led_right

goto ir_cal

end
```

PROGRAM 5.4

ircal-serpent.hex file
listing

```
:10000000012883168501033086008312051183116316AB
:100010000511831285118316851183126400861  8D9
:10002000192805158316051183128515831685116  8
:10003000831205118316051183128511831685110C
:0800400083120E286300222840
:02400E00F53F7C
:00000001FF
```

Motion Control

The next task will be to coordinate the movement of each of the snake's body segments to achieve locomotion. To produce a forward movement, our snake will move its body in a sine wave pattern vertically, with a slight side to side movement of the horizontal segments. The use of servos makes this sort of programming easy because all that is needed to coordinate this pattern is to give each of the servos two sets of movement positions. The body segments will move through the complete range of motion between the two sets of points determined by the position values. This means that we really only need to set the servo positions for all of the servos twice, and then repeat the pattern to get the snake to move forward. The same holds true when sequencing the servos and body segments for a left or right turning movement. **Figure 5.59** shows the pulsout values for the extreme and middle positions, along with the microcontroller port address for each servo. This information will be needed when putting the control program together.

To sequence the forward movement of the snake, a sine wave pattern can be generated by using the servo position values shown in **Table 5.7**. The servos that move the horizontal body segments also move in a slight side to side movement to aid in locomotion. **Figure 5.60** shows the sequence that the snake's body goes through when moving in a forward direction. Frame number 1 shows the snake resting before the sequence begins. Frame number 2 shows the body segment positions that correspond to the first set of positions in **Table 5.7**. Frame number 3 shows that the snake's body moves through the original position on its way to the

FIGURE 5.59

Microcontroller port addresses for each of the body segment servos.

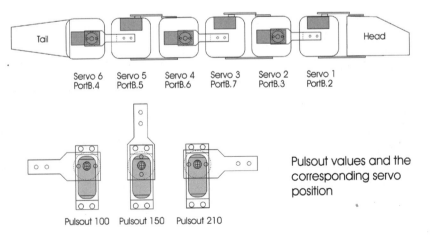

| Servo 6 | Servo 5 | Servo 4 | Servo 3 | Servo 2 | Servo 1 |
| PortB.4 | PortB.5 | PortB.6 | PortB.7 | PortB.3 | PortB.2 |

Pulsout 100   Pulsout 150   Pulsout 210

Pulsout values and the corresponding servo position

second set of positions in **Table** 5.7. Frame number 4 shows the body segment positions that correspond to the second set of positions in **Table** 5.7. When the sequence is running, the body moves in a sine wave pattern. For the snake to continue moving forward, this entire sequence repeats. In the control program, the servo positions only need to be set twice, and then the sequence repeats. If you wish to experiment, you could program sequences with more intermediate positions for a smoother sine wave.

Body Position 1

Servo and port address	Pulsout value
1—PortB.2	157
2—PortB.3	210
3—PortB.7	143
4—PortB.6	100
5—PortB.5	157
6—PortB.4	210

Body Position 2

Servo and port address	Pulsout value
1—PortB.2	143
2—PortB.3	100
3—PortB.7	157
4—PortB.6	210
5—PortB.5	143
6—PortB.4	100

TABLE 5.7

Servo Position Values to Sequence Forward Movement of the Snake

FIGURE 5.60

Sequence of body
positions during forward
movement.

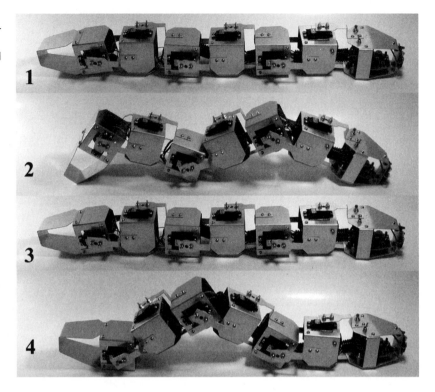

To make the robot snake turn to the left, the same sine wave pattern will need to occur in the vertical moving body segments, but the snake's body will also need to oscillate between the middle position and a position where the body is arched to the left. The pulsout values needed to control this movement are listed in **Table 5.8**, and will be used when programming the snake. **Figure 5.61** shows the two positions that the snake's body will oscillate between to make a turn to the left. I found that the snake has the ability to turn to the left or right much faster than it can travel in the forward direction. Although a side-winding routine will not be covered in this chapter, with enough experimentation, the snake can be made to side-wind as its primary mode of locomotion. When the snake is traveling forward and then moves quickly into a turn, the effect is quite surprising and very lifelike.

Body Position 1

Servo and port address	Pulsout value
1—PortB.2	150
2—PortB.3	210
3—PortB.7	150
4—PortB.6	100
5—PortB.5	150
6—PortB.4	210

Body Position 2

Servo and port address	Pulsout value
1—PortB.2	100
2—PortB.3	100
3—PortB.7	100
4—PortB.6	210
5—PortB.5	100
6—PortB.4	100

TABLE 5.8

Servo Position Values Needed to Sequence a Left Turn

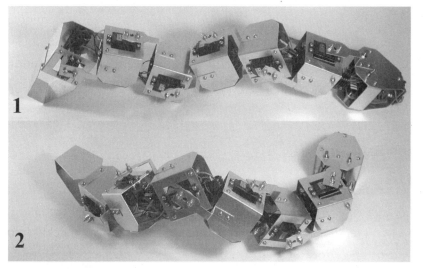

FIGURE 5.61

Sequence of body positions during a left turn.

To make the robot snake turn to the right, the same sine wave pattern will need to occur in the vertical moving body segments, but the snake's body will also need to oscillate between the middle position and a position where the body is arched to the right. The pulsout values needed to control this movement are listed in **Table 5.9** and will be used when programming the snake. **Figure 5.62** shows the two positions that the snake's body will oscillate between to turn to the right. You might have noticed that when positioning the robot's body to the right, smaller pulsout values were used. This is to take into account the extra weight of the servos that are positioned on the right side of the snake's body.

TABLE 5.9

Servo Position Values Needed to Sequence a Right Turn

Body Position 1

Servo and port address	Pulsout value
1—PortB.2	150
2—PortB.3	210
3—PortB.7	150
4—PortB.6	100
5—PortB.5	150
6—PortB.4	210

Body Position 2

Servo and port address	Pulsout value
1—PortB.2	190
2—PortB.3	100
3—PortB.7	190
4—PortB.6	210
5—PortB.5	190
6—PortB.4	100

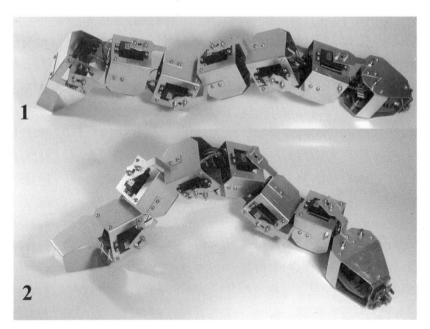

FIGURE 5.62

Sequence of body positions during a right turn.

Infrared Sensor

The next section outlines conditioning the input received by the infrared sensor. The motion control algorithms and sensor input routines will then be put together into one main control program.

The infrared software routine will need to take input from the infrared sensor so that the robot can change its behavior to safely avoid any obstacles it may encounter while moving through its environment. A software subroutine will be developed to monitor the infrared sensor modules, perform signal processing to clean up any background noise or transient signals to make the information more useful, and then return results to the robot's main program. In this behavior-based method of artificial intelligence, the robot will continue on with the dominant behavior of exploring, and will change that course of action immediately based on sensor input.

We want the main program to call the subroutine and have the subroutine simply return a value of either a 1 or a 0, with 0 indicating that no object was sensed and 1 indicating that an object is present. These values will be stored in the variable object_detect. When the program execution is returned back to the main program, certain decisions can easily be made, based on this information.

The infrared subroutine takes 40 samples from the module and counts the number of positive hits received. The number of samples taken can also be configured by changing the variable num_samples. Because of stray infrared and signals from the environment, the module is constantly producing false positive signals that are referred to as "noise." The average acceptable amount of noise picked up by the sensor module is called the *noise floor*. The routine needs to set a threshold point above the typical amount of noise and report a sensed object only if the number of positive signals received throughout the number of samples taken exceeds the noise floor.

With the PNA4602M sensor modules, I found that the typical false positive was actually very low—five for every 40 samples taken. To be on the safe side, the threshold is set at 25 for every 40 samples, to ensure that an object is present. By changing the threshold value, you can change the sensitivity and distance detection response of the module. If you want a more accurate reading, the num_samples value can be increased, but will take more time for the routine to execute.

The last option is using the mode select push button to invoke the infrared sensor calibration routine. This will enable the user to simply push the button on the robot's head to calibrate the sensor, as described earlier. The experimenter can also develop a software routine to use the push button to choose different modes of behavior when the robot starts up. When the main software routine senses that the button has been pushed, it goes into a tight loop until it senses that the switch has been let up before going to the infrared calibration routine. This is so that when the program execution jumps to the calibration routine, it does not immediately jump back to the main routine because the operator still has the button pushed.

The main robot snake control program is called serpentronic.bas and is listed in **Program** 5.5. The program operates by constantly moving the snake in a forward direction, monitoring the infrared sensor and then responding by turning either left or right if an obstacle was sensed. Compile serpentronic.bas and then program the PIC 16F84 with the serpentronic.hex file listed in **Program** 5.6. The program can be put into the infrared calibration mode by holding down the push button.

PROGRAM 5.5

serpentronic.bas
program listing

```
'_____
'
' Name       : Serpentronic.bas
' Compiler   : PicBasic Pro - MicroEngineering Labs
' Notes      : Complete control Program for the robot
'            : snake. Mode select push-button switch
'            : allows the infrared sensor to be easily
'            : calibrated. The robot will stop and turn
'            : if an obstacle is encountered.
'_____


' PortA set as outputs
trisa = %00000000

' PortB set as outputs. pins 0-1 inputs
trisb = %00000011


'_____

' initialize variables

led_left          VAR PORTA.2
led_right         VAR PORTA.3
piezo             VAR PORTA.4

cal_switch        VAR PORTB.0
```

PROGRAM 5.5

serpentronic.bas
program listing
(continued)

```
ir_input        VAR PORTB.1
servo_1         VAR PORTB.2
servo_2         VAR PORTB.3
servo_3         VAR PORTB.7
servo_4         VAR PORTB.6
servo_5         VAR PORTB.5
servo_6         VAR PORTB.4


ir_count        VAR byte
temp            VAR BYTE
object_detect   VAR BYTE
num_samples     VAR Byte
threshold       VAR BYTE
rand            VAR WORD
timer           VAR BYTE
temp1           VAR BYTE
i               VAR BYTE

look_right      VAR BYTE
look_left       VAR BYTE
turn_count      VAR BYTE

servo1          VAR BYTE
servo2          VAR BYTE
servo3          VAR BYTE
servo4          VAR BYTE
servo5          VAR BYTE
servo6          VAR BYTE

low led_left
low led_right

Low servo1
Low servo2
Low servo3
Low servo4
Low servo5
```

Low servo6

turn_count = 0
num_samples = 40
threshold = 25

PROGRAM 5.5

serpentronic.bas
program listing
(continued)

```
'_____

' create randon noises and flash LED's

      For temp1 = 1 to 5
         High led_left
         Low led_right
         GoSub randomize
         Pause 50

         Low led_left
         High led_right
         GoSub randomize
         Pause 50
      Next temp1

      Low led_right

'_____

' start main execution

start:

      If cal_switch = 1 then
         pause 50
         release_calibrate:
         If cal_switch = 1 then
            goto release_calibrate
         else
            Sound piezo,[120,4,90,2,100,2,110,4]
            pause 50
            goto ir_cal
         endif
```

PROGRAM 5.5

serpentronic.bas
program listing
(continued)

```
        endif

gosub infrared

    if object_detect = 1 then
        high led_left
        high led_right
        Sound piezo,[100,4,90,2]
        servo1 = 180
        gosub servo
        servo1 = 120
        gosub servo
        turn_count = turn_count + 1
          if turn_count.0 = 1 then
            gosub slide_right
          else
            gosub slide_left
          endif

    endif

        low led_left
        low led_right

gosub forward

goto start

'Subroutines start here

'_____

' slither forward routine in a sine wave pattern

forward:
    servo1 = 157
    servo2 = 210
    servo3 = 143
    servo4 = 100
```

```
    servo5 = 157
    servo6 = 210
    GoSub servo
    servo1 = 143
    servo2 = 100
    servo3 = 157
    servo4 = 210
    servo5 = 143
    servo6 = 100
    GoSub servo
return
```

```
'_____

' right turn movement routine

slide_right:

For temp1 = 1 to 3
    servo1 = 150
    servo2 = 210
    servo3 = 150
    servo4 = 100
    servo5 = 150
    servo6 = 210
    GoSub servo
    servo1 = 190
    servo2 = 100
    servo3 = 190
    servo4 = 210
    servo5 = 190
    servo6 = 100
    GoSub servo
next temp1
return


'_____

' left turn movement routine
```

PROGRAM 5.5

serpentronic.bas
program listing
(continued)

```
slide_left:

For temp1 = 1 to 3
    servo1 = 150
    servo2 = 210
    servo3 = 150
    servo4 = 100
    servo5 = 150
    servo6 = 210
    GoSub servo
    servo1 = 100
    servo2 = 100
    servo3 = 100
    servo4 = 210
    servo5 = 100
    servo6 = 100
    GoSub servo
Next temp1
return


'_____

' random sound generator subroutine

randomize:

Random rand
i = rand & 31 + 64
Sound piezo,[i,4]
Return


'_____

' infrared detection subroutine

infrared:

    ir_count = 0
    object_detect = 0
```

```
    for temp = 1 to num_samples
        if ir_input = 0 then
        ir_count = ir_count + 1
        endif
    next

    if ir_count >= threshold then
    object_detect = 1
    endif

return


'_____

' subroutine to calibrate I.R. sensors

ir_cal:

    If ir_input = 0 then
        high led_left
        high led_right
    endif
        low led_left
        low led_right

    If cal_switch = 1 then
        pause 50
        button_release:
        If cal_switch = 1 then
            goto button_release
        else
            Sound piezo,[120,4,90,2,100,2,110,4]
            pause 50
            goto start
        endif
    endif

goto ir_cal
```

PROGRAM 5.5

serpentronic.bas
program listing
(continued)

PROGRAM 5.5

serpentronic.bas
program listing
(continued)

```
'_____

' subroutine to set servos

servo:

        For timer = 1 to 20
        PulsOut servo_1,servo1
        PulsOut servo_2,servo2
        PulsOut servo_3,servo3
        PulsOut servo_4,servo4
        PulsOut servo_5,servo5
        PulsOut servo_6,servo6
        Pause 12
        Next timer


Return
```

PROGRAM 5.6

serpentronic.hex file
listing

```
:100000009128A0003F200C080D0403198C2886209D
:100010008413200880066464000D280E288C0A03191A
:100020008D0F0B2880068C288F0022088400200961
:10003000402084138F0803198C28F03091000E089B
:100040080389000F03091030319910003198F0359
:1000500003198C282B28552003010C1820088E1F0B
:1000600020088E0803190301900F382880061F28E6
:100070003928000022284320FF3A80054028FF3A13
:10008000841780058C28940006309419053084006C
:100090000308A001408073982070134023404341E
:1000A000083410342034403480340D080C04031913
:1000B0008C0A80300C1A8D060C198D068C188D0652
:1000C0000D0D8C0D8D0D8C288F018E00FF308E074D
:1000D000031C8F07031C8C2803308D00DF30722037
:1000E00066288D01E83E8C008D09FC30031C7B28BE
:1000F0008C07031878288C0764008D0F78280C185B
:100100081288C1C85280000852808008C098D0911
:100110008C0A03198D0A08008313031383126400E9
:100120008008316850103308600831205118316AB
:100130000511831285118316851183122C083B202B
```

```
:100140002D083B202E083B202F083B2030083B2069
:1001500031083B20B6012830AA001930B400013024
:10016000B3006400063033020318CE280515831649
:100170000511831285118316851183128721323070
:1001800064200511831605118312851583168511C8
:10019000831287213230642OB30FB1288511831672
:1001A000851183126400061CF32832306420640039
:1001B000061CDC28D728F3280530A2001030A00048
:1001C00078308E00043014205A308E000230142013
:1001D00064308E00023014206E308E000430142003
:1001E00032306420B3299C2164002B08013C031D9C
:1001F0001A2905158316051183128515831685119 5
:100200000530 8312A2001030A00064308E0004304C
:1002100014205A308E0002301420B430AC00E82193
:100220007830AC00E821B60A6400361C19293F2159
:100230001A29632105118316051183128511 83166E
:10024000851183122421D2289D30AC00D230AD001C
:100250008F30AE006430AF009D30B000D230B100BE
:10026000E8218F30AC006430AD009D30AE00D2305C
:10027000AF008F30B0006430B100E82108000130D9
:10028000B300640004303302031862299630AC00D6
:10029000D230AD009630AE006430AF009630B00082
:1002A000D230B100E821BE30AC006430AD00BE30C9
:1002B000AE00D230AF00BE30B0006430B100E821F3
:1002C000B30F412908000130B30064000430330249
:1002D00003188 6299630AC00D230AD009630AE00BF
:1002E0006430AF009630B000D230B100E821643005
:1002F000AC006430AD006430AE00D230AF0064308A
:10030000B0006430B100E821B30F6529080024086B
:100310008C0025088D0055200C08A4000D08A500B0
:100320005F302405A6000530A2001030A00026088A
:100330008E00043014200800A701AB010130B20088
:10034000640032082A02031CAB2964008618A9291C
:10035000A70AB20FA029640034082702031CB2299F
:100360000130AB00080064008618BE29051583160D
:100370000511831285158316851183120511831 6C5
:100380000511831285118316851164008312061CE2
:10039000E729323064206400061CD029CB29E729E4
```

PROGRAM 5.6

serpentronic.hex file
listing (continued)

PROGRAM 5.6

serpentronic.hex file
listing (continued)

```
:1003A0000530A2001030A00078308E0004301420F8
:1003B0005A308E000230142064308E000230142037
:1003C0006E308E00043014203230642OD228B329DD
:1003D0000130B50064001530350203181D2A2C08C1
:1003E0008C008D0106308400043001202D088C0023
:1003F0008D0106308400083001202E088C008D010C
:10040000630840080300120 2F088C008D010630DA
:1004100084004030012030088C008D0106308400BB
:100420002030012031088C008D010630840010300E
:0C04300001200C306420B50FEA29080000
:02400E00F53F7C
:00000001FF
```

Summary

This concludes the construction and programming of the robot
snake. Much more can be done with this robot than what has been
covered. A remote control can easily be added to this project, since
there are two connectors on the controller board for this purpose.
(Chapter 12 of the first book in this series, *Insectronics*, has details.)

Other customizations that can be added are:

- Use the infrared sensor and the snake's head movement to
  scan the area around the snake for objects. Use this information to determine the correct path before moving.

- Create a skin for the robot using a waterproof material such
  as latex rubber.

- Add a wireless video camera.

- Develop a side-winding movement routine.

- Figure out a routine that will enable the robot to move in
  reverse, unlike a real snake.

- Add a tilt sensor so that the robot will know when it has tipped over, and can then right itself.

- Write a routine enabling the snake to roll over.

To see movies of the snake in action, go to the author's Web site located at www.thinkbotics.com.

6

Crocobot: Build Your Own Robotic Crocodile

Crocodilians

Crocodiles, alligators, and gharials are all part of a group of reptiles known as the crocodilians. The bodies of animals in this group are covered in a tough, leathery skin that is strengthened with plates known as *osteoderms*, or bone skin. Crocodilians are unable to sweat through their tough skin. They keep themselves cool by resting with their mouths open, permitting moisture to evaporate from the mucous membranes. Although modern crocodilians have an almost primeval appearance, they are actually quite advanced, possessing an elaborate, four-chambered heart similar to that of a mammal. It is generally accepted by biologists that birds, rather than other reptiles, are the nearest living relatives of modern crocodilians. All crocodilian species, except for the American alligator, are endangered in at least part of their ranges, and some are threatened with extinction as a result of habitat destruction, hunting, or pollution.

Crocodiles and their method of locomotion are the inspiration for the robot in this chapter. **Figure 6.1** shows the Nile crocodile along with its biologically inspired robotic counterpart. The robot croc-

FIGURE 6.1

A crocodile and its biologically inspired counterpart.

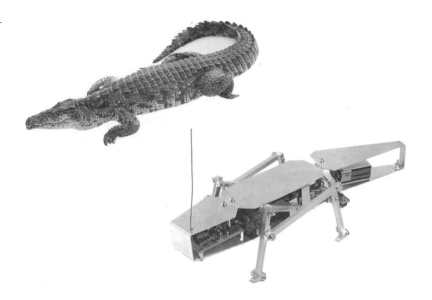

odile measures 14 inches in length from head to tail, and is 5 inches wide.

Moving the body from one location to another is one of the most important everyday tasks for animals. They must be able to move from place to place during the activities necessary for survival. These activities include thermoregulation, finding food, social interactions, nesting, and escape from threats. While crocodiles spend much of the day motionless or moving very little, it is a mistake to think that they are not very active. Crocodiles are capable of moving at surprising speed when required. Crocodiles have three basic styles of moving on land. These methods of locomotion are usually referred to as the belly crawl, the high walk, and the gallop. The belly crawl is very similar in form to the way that a lizard moves. The legs are splayed out to the sides and the center of gravity is low. The belly crawl is used on land and very shallow water. The crocodile uses its front and hind limbs to achieve locomotion. The crocodile's whole body and tail undulates rapidly from side to side when walking. The belly crawl is probably the most commonly used way in which crocodiles move around on land. It is usually slow, although it can be modified so that the crocodile reaches speeds of 5 to 10 kilometers per hour when required. Although the term "belly crawl" implies a certain style of locomotion, in reality there are several variations on this gait suited to different situations, and only at very slow speeds does the crocodile actually crawl, as the name suggests.

The high walk and gallop are unlike a reptilian gait. The crocodile walks more like a mammal during the high walk. The gallop is very spectacular to watch, and propels even large crocodiles away from potential danger at very high speeds. The robotic crocodile in this chapter will use a method of walking on four legs where the body is raised completely above the ground.

Overview of the Crocobot Project

The robot crocodile that will be built and programmed in this chapter is controlled remotely by a human operator via a wireless data link. The robot and the remote control that will be built are shown in **Figure 6.2**. The wireless data is transmitted from the controller and received by the robot using RF modules built by a company called Linx Technologies. The robot achieves locomotion using four legs that are driven by a twin-motor gearbox. The geared motors operate on voltages between 3 and 6 volts, making them perfect for small walking robots. The motors are controlled using the L298 dual full-bridge driver. The motor driver takes its control signals from a PIC 16F84 microcontroller. The microcontroller will also be used to interpret the control commands sent from the hand held remote control. The remote control uses a PIC 16C71 microcontroller featuring four analog to digital converters. Two of the analog to digital converters will be used to monitor the position of the control stick on the remote control device. This is accomplished by reading the voltages produced by the poten-

FIGURE 6.2

Crocobot with remote
control device.

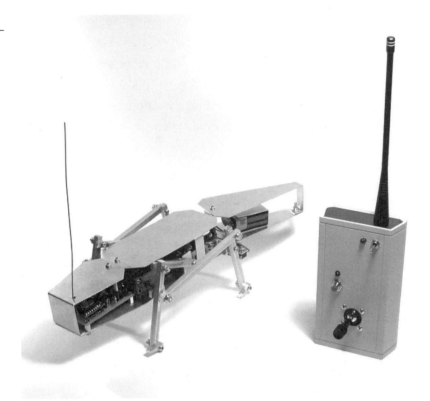

tiometers attached to the X and Y axis. When the position of the
control stick is determined, certain control information is transmitted to the robot. Because a wireless data link is being used to
remotely control the robot, the experimenter is not limited to a
certain number of control channels, as are imposed when a regu-
lar model airplane remote control system is used. The experimenter has the option of adding any number of other devices.

Mechanical Construction of Crocobot

The construction of the robot crocodile will begin with the
mechanical construction of the body, head, and tail. The parts
needed for the mechanical construction are listed in **Table 6.1**.

Parts	Quantity	TABLE 6.1
1/16-inch thick aluminum stock	8-foot x 10-foot piece	Parts List for Crocobot's Mechanical Construction
1/4-inch × 1/4-inch aluminum stock	34 inches	
6/32 × 1/2-inch machine screws	32	
6/32 × 1-inch machine screws	2	
6/32 locking nuts	34	
6/32 nylon washers	14	
Tamiya twin motor gear box	1	
Connector wire	9 feet	
Heat-shrink tubing	2 inches	
4-post female header connector	3	

The body, head, and tail are constructed using 1/16-inch flat aluminum.

The construction of the robot crocodile will start with the assembly of the Tamiya twin motor gearbox. It is available from a hobby robotics supplier called HVW Tech, and can be purchased from its Web site located at www.hvwtech.com. The gearbox is sold as a kit and needs to be assembled before it can be used. **Figure 6.3** shows the entire kit before assembly.

FIGURE 6.3

Tamiya twin motor
gearbox before
assembly.

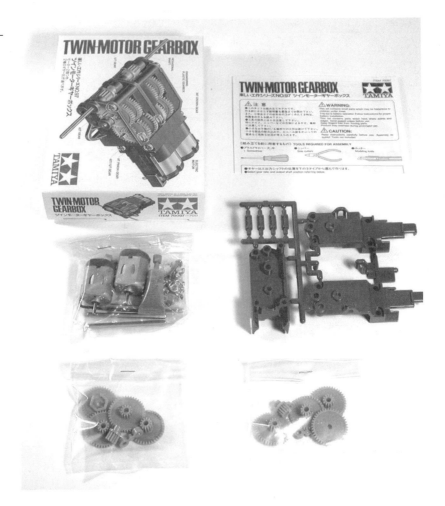

Assembling the twin motor gearbox. Take all of the parts out of the box and unfold the instruction sheet. The gearbox has two possible configuration options of standard speed with a gear ratio of 58:1, or low speed with a gear ratio of 203:1. The gearbox will be assembled for use with the low speed option. The first thing that needs to be done when assembling the gearbox is to position a gear hub on each of the two hexagonal output shafts, as shown in **Figure 6.4**. Thread a grub screw into each of the gear hubs with the hex wrench that was supplied with the kit. Use piece M3 to set the proper position of the gear hubs, and then tighten in place with the hex wrench.

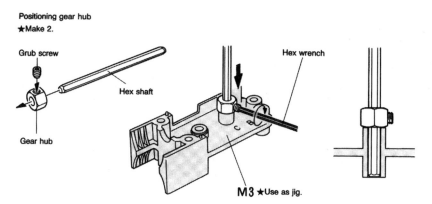

FIGURE 6.4

Procedure to attach gear hubs to the hexagonal output shafts.

Break apart each of the gearbox body sections and plastic spacers from the injection-molded piece, and trim off any rough edges with a small knife. Locate the gears, eyelets, screws, and output shafts, then assemble according to **Figure 6.5**.

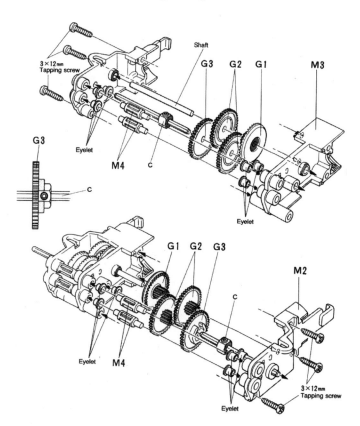

FIGURE 6.5

Gearbox assembly diagram.

FIGURE 6.6

Installing motors in the gearbox.

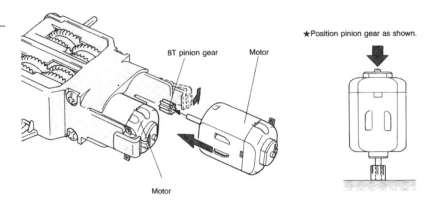

8T pinion gear   Motor

★Position pinion gear as shown.

Motor

Place a pinion gear onto the end of each motor shaft so that the end of the shaft is level with the end of the gear. Install each motor in the gearbox by sliding it into place, as shown in **Figure 6.6**. The plastic clips on the gearbox body will snap into place and secure the motors in position.

When the gearbox is complete, mark each shaft at 5/8 of an inch from the body and cut with a hacksaw. The finished gearbox, ready for use with Crocobot, is shown in **Figure 6.7**.

FIGURE 6.7

Completed twin motor gearbox with a 203:1 gear reduction.

Cut each output shaft to a length of 5/8-inch

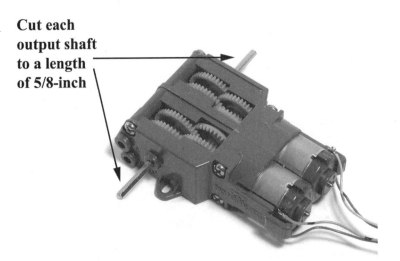

Constructing the Chassis

The main body chassis is constructed using a piece of 1/16-inch thick flat aluminum, and is labeled as part A. Cut a piece to the size of 9 inches in length by 2-1/2 inches in width. Use **Figure 6.8** as a cutting and drilling guide.

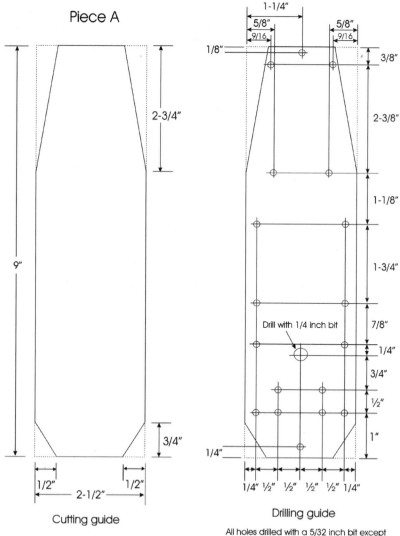

FIGURE 6.8

Cutting and drilling guide for main body chassis.

Piece A

Cutting guide

Drilling guide

All holes drilled with a 5/32 inch bit except where marked.

FIGURE 6.9

Cutting, bending, and drilling guide for mounting brackets.

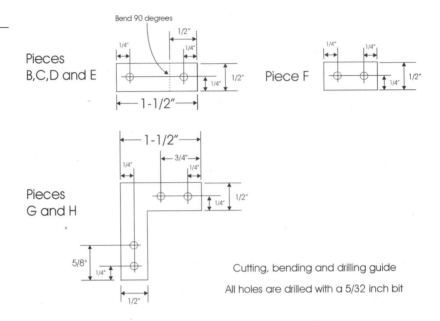

Cutting, bending and drilling guide

All holes are drilled with a 5/32 inch bit

Fabricate four leg support brackets using the 1/16-inch aluminum, as detailed in **Figure 6.9**. These pieces are identified by the letters B, C, D, and E. When the pieces are finished, use a file to remove any rough edges.

Create a single support bracket according to the dimensions shown in **Figure 6.9**. This part is labeled piece F, and is also constructed using the 1/16-inch aluminum. Fabricate two L-shaped limit switch mounting brackets identified as pieces G and H in **Figure 6.9**, also using the 1/16-inch aluminum.

Attach the leg mounting brackets (pieces B, C, D, and E) to the body chassis (piece A) using four 6/32-inch × 1/2-inch machine screws and locking nuts, as shown in **Figure 6.10**. Note that when pieces B and C are mounted, the 1/4-inch side of each bracket is attached to the chassis, and when pieces D and E are mounted, the 1-inch side of each bracket is attached to the chassis. **Figure 6.10** shows the mounting brackets attached to the robot chassis.

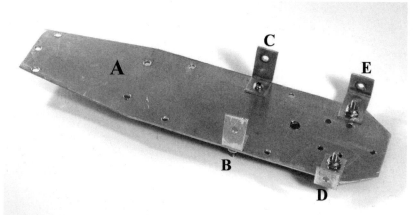

FIGURE 6.10

Leg mounting brackets attached to the robot chassis.

Cut four pieces of connector wire to a length of 6-1/2 inches each. Cut four pieces of heat-shrink insulator tubing to a length of 1/4-inch each. Use **Figure 6.11** to attach the motor to a 4-connector female header using the connector wire. Use the heat-shrink tubing to protect from shorts at the header.

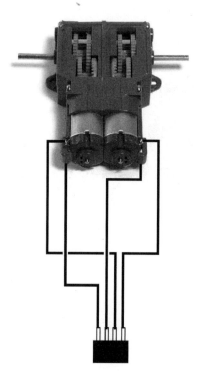

FIGURE 6.11

Motor wiring diagram.

When the motor has been wired to the header, attach it to the robot chassis with the two mounting nuts and bolts that came with the motor kit. **Figure 6.12** shows the position of the motor mounted to the chassis.

Constructing the Body Covers and Tail Section

The next step will be to construct the robot's top body cover. The body cover is made up of three parts and will also carry three AA battery holders and batteries that are used as the power supply for the direct current motors. Cut a piece of the 1/16-inch thick aluminum to a size of 2-1/4 inches by 4-1/4 inches. Use **Figure 6.13** as a cutting, drilling, and bending guide. This piece is the robot's head cover piece, and is identified with the letter I.

Piece I

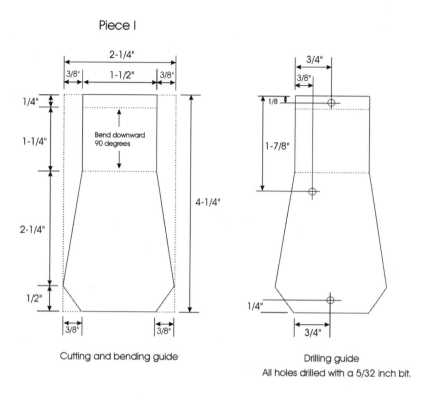

FIGURE 6.13

Cutting, drilling, and bending diagram for head cover piece.

Cutting and bending guide

Drilling guide
All holes drilled with a 5/32 inch bit.

Cut another piece of the 1/6-inch thick aluminum to a size of 5-1/2 inches by 6 inches. Cut, drill, and bend the piece, as shown in **Figure 6.14**. This piece will make up the body cover, and is attached to the head cover piece. This piece is identified as J.

Locate mounting bracket (F) and use it to join the head cover piece (I) to the body cover piece (J) using two 6/32-inch × 1/2-inch machine screws and locking nuts, as shown in **Figure 6.15**. When the two pieces are joined, wire three single AA battery holders in series, as shown in **Figure 6.16**, so that a total of 4.5 volts are produced. Solder the negative and positive outputs to a 2-connector male header. Use a glue gun to glue the battery holders to the top body cover in the position, as shown in **Figure 6.15**. **Figure 6.17** shows the completed top body cover from the top view.

FIGURE 6.14

Cutting, drilling, and bending diagram for body cover piece.

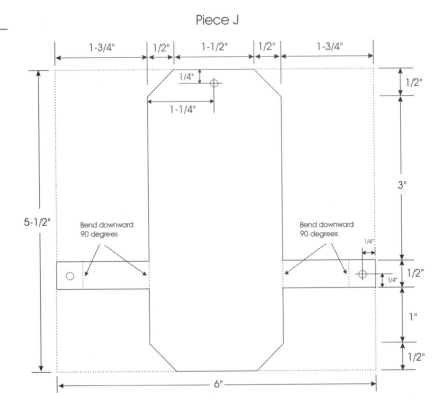

Piece J

FIGURE 6.15

Pieces I and J attached with mounting bracket—underside view.

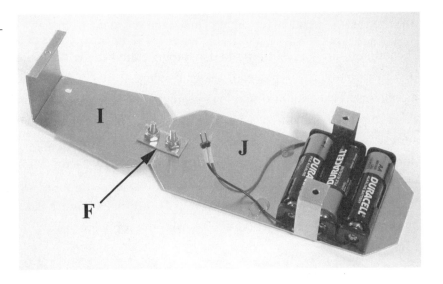

4.5 volts

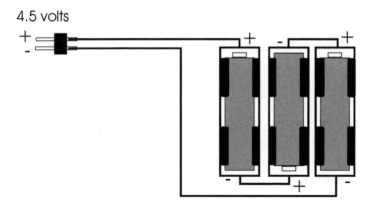

FIGURE 6.16

Motor power supply wiring diagram.

FIGURE 6.17

Completed top body cover—top view.

The robot's tail section will be added to the end of the chassis and will contain the 9-volt battery holder and 9-volt battery. The tail will swing from side to side as the robot walks, adding to the reptilian realism of the robot. To construct the tail section, cut two pieces of 1/16-inch thick aluminum to a size of 6-3/4 inches by 2-1/4 inches. Use the diagrams in **Figure 6.18** (piece K) and **Figure 6.19** (piece L) to cut, drill, and bend the pieces. Construct the 9-volt battery holder (piece M) using 1/16-inch thick aluminum, as detailed in **Figure 6.20**. Assemble each of the pieces, as shown in **Figure 6.21**, using three 6/32-inch × 1/2-inch machine screws and locking nuts.

FIGURE 6.18

Upper tail section cutting, drilling, and bending diagram.

Piece K

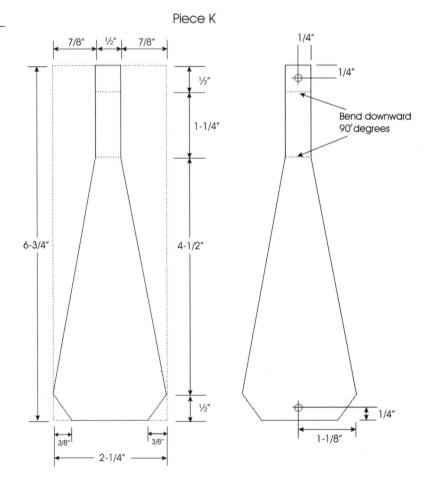

Piece L

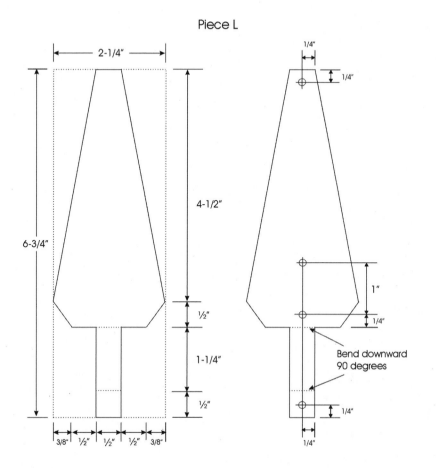

FIGURE 6.19

Lower tail section cutting, drilling, and bending diagram.

Bend downward 90 degrees

FIGURE 6.20

Battery holder cutting, drilling, and bending guide.

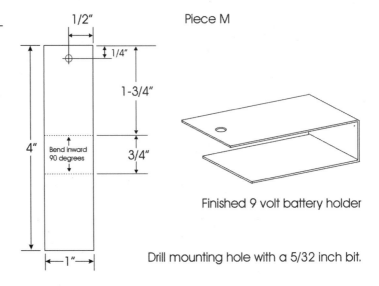

Piece M

1/2"

1/4"

1-3/4"

4"

Bend inward 90 degrees

3/4"

1"

Finished 9 volt battery holder

Drill mounting hole with a 5/32 inch bit.

FIGURE 6.21

Completed tail section with battery holder.

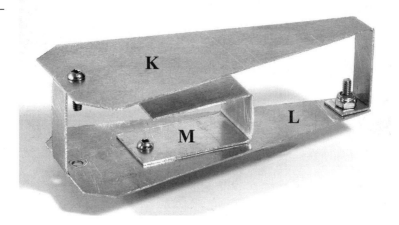

Wiring the Limit Switches

Mount the leg limit switches to the mounting brackets labeled G and H with appropriately sized machine screws and nuts, oriented as shown in **Figure 6.22**. Cut four wires to a length of 8 inches. Solder two of these wires to a 2-connector female header. Locate another 2-connector female header and solder the other two wires to it. Protect each of the connections with a 1/4-inch length of heat-shrink tubing. Cut two more wires to a length of 5 inches. Wire up the leg limit switches, as shown in **Figure 6.22**. The finished leg limit switches with connectors and mounting brackets are shown in **Figure 6.23**. Attach the mounting brackets with the limit switches to the bottom of the chassis using four 6/32-inch × 1/2-inch machine screws and locking nuts, as shown in **Figure 6.24**.

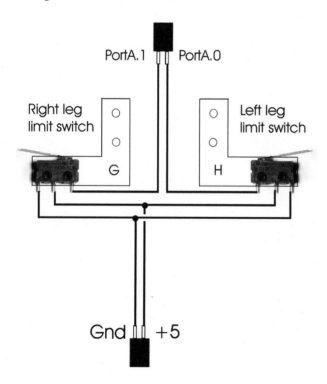

FIGURE 6.22

Limit switch wiring diagram.

FIGURE 6.23

Completed limit
switches wired and
attached to mounting
brackets.

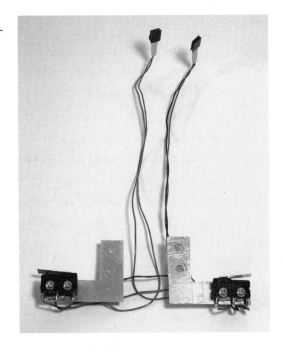

FIGURE 6.24

Limit switches and
mounting brackets
attached to chassis.

Constructing the Legs

The legs, feet, and motor shaft mounts will be created using 1/4-inch × 1/4-inch aluminum square stock. Start by fabricating two motor output shaft mounts according to the dimensions shown in **Figure 6.25**. The two parts are identified as N and O. When the pieces are finished, thread a 6/32-inch × 1/4-inch machine screw into the hole that has been threaded with a 6/32-inch tap. **Figure 6.26** shows a completed motor shaft mount.

Pieces N and O

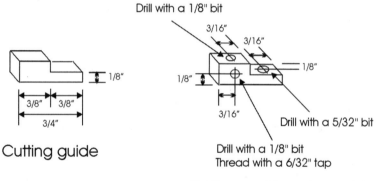

Drill with a 1/8" bit

3/16″

3/16″

1/8″   1/8″   1/8″

3/16″

3/8″   3/8″   3/16″

3/4″   Drill with a 5/32" bit

Cutting guide   Drill with a 1/8" bit
Thread with a 6/32" tap

Drilling and tapping guide

FIGURE 6.25

Motor output shaft mount fabrication diagram.

FIGURE 6.26

Completed motor output shaft mount.

Using the 1/4-inch × 1/4-inch aluminum stock, fabricate the four leg pieces (P, Q, R, and S) and the mechanical linkage pieces (T and U), as detailed in **Figure 6.27**. Construct the mechanical linkage pieces V and W and the four feet (X1, X2, X3, and X4), as outlined in **Figure 6.28**. When all of these pieces are complete, the robot's legs will be assembled.

FIGURE 6.27

Leg and mechanical linkage construction diagram.

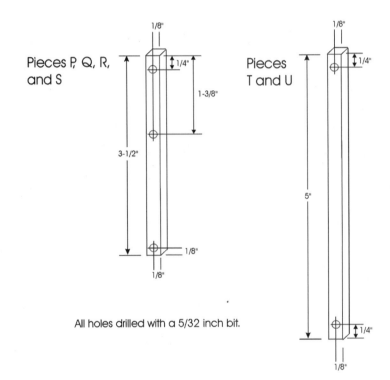

Pieces P, Q, R, and S

Pieces T and U

All holes drilled with a 5/32 inch bit.

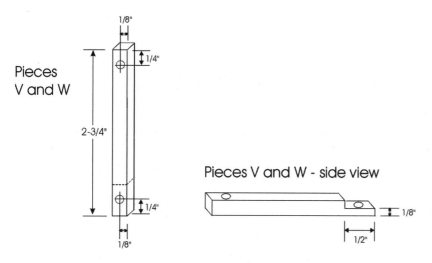

FIGURE 6.28

Feet and mechanical linkage construction diagram.

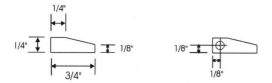

Cutting and drilling guide

All holes drilled with a 5/32 inch bit.

Assembling the Legs

Start by connecting the motor shaft mounts (pieces N and O) to the motor shafts so that the motor shafts are flush with the outer sides of the mounts when they are placed. Tighten the screw on the mounts so that each mount is secure on the motor's hex shafts. Use **Figure 6.29** and **Figure 6.30** as a guide to assembling the legs. Note that the leg pieces attached to the motor shaft mounts use 6/32-inch × 1-inch machine screws and locking nuts. All of the others use 6/32-inch × 3/4-inch machine screws and locking nuts. The foot piece machine screws and locking nuts should be as tight as possible. All of the other joints should have a 6/32 nylon washer between metal pieces, and the locking nuts

FIGURE 6.29

Leg parts placement for the robot's left side.

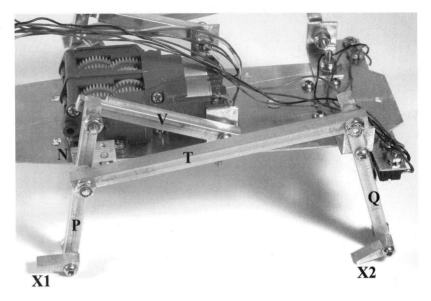

FIGURE 6.30

Leg mechanism parts placement.

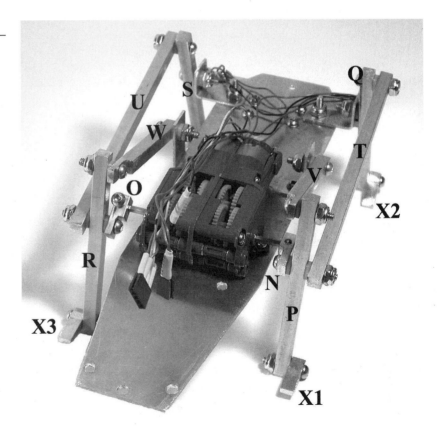

should be fastened with just enough pressure to allow the parts to move freely without any resistance.

Cut six connector wires to a length of 6 inches each. Wire the power switch, 9-volt battery strap, and three female header connectors, as indicated in **Figure 6.31**. When the switch and connectors are finished, mount the switch in the 1/4-inch hole in the robot chassis with the switch mechanism facing down toward the bottom of the robot, and the 9-volt battery strap facing toward the back. Now that the mechanical and electrical systems are in place, the next step is to add the electronics.

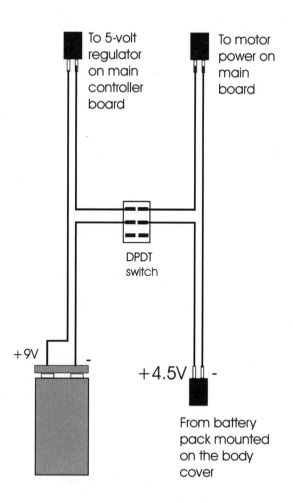

To 5-volt regulator on main controller board

To motor power on main board

DPDT switch

+9V

+4.5V

From battery pack mounted on the body cover

FIGURE 6.31

Power switch wiring diagram.

The Controller Circuit Board

The robot's main controller will integrate a PIC 16F84 microcontroller, a Lynx radio receiver module, and an L298 dual motor controller chip all on a 1-1/2 inch by 2-1/2 inch circuit board. The schematic for the controller board is shown in **Figure 6.32**.

FIGURE 6.32

Crocobot's main controller board.

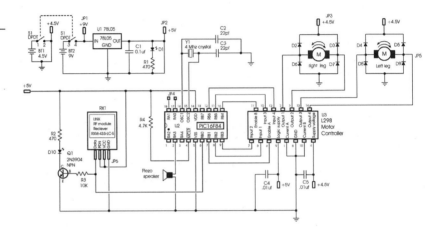

The PIC 16F84 microcontroller is used to interpret the serial information that is received from the Lynx radio receiver module, monitor the leg limit switches, and control the motors via the L298 motor controller I.C. The 16F84 microcontroller is clocked at 4 MHz and operates from a 5-volt direct current (DC) supply that is produced from a 78L05 voltage regulator, with the source being a 9-volt battery in the robot's tail section. The motors operate from their own 4.5-volt supply contained in the robot's top cover. Six of the PIC 16F84 port B pins will be connected to the L298 to control the motors. The parts necessary to construct the main board are listed in **Table 6.2**.

Part	Quantity	Description		
Semiconductors			**TABLE 6.2**	
U1	1	78L05 5V regulator	Parts List for Crocobot's Main Controller Board	
U2	1	PIC 16F84 flash microcontroller mounted in socket		
U3	1	L298 dual full-bridge driver		
RX1	1	Lynx RXM-433-LC-S RF receiver module		
D1	1	Red light-emitting diode		
D2—D9	8	Diodes 1N4001		
D10	1	Green light-emitting diode		
Q1	1	2N3904 NPN transistor		
Resistors				
R1, R2	2	470 Ω 1/4-watt resistor		
R3	1	10 KΩ 1/4-watt resistor		
R4	1	4.7 KΩ 1/4-watt resistor		
Capacitors				
C1	1	0.1 µf		
C2, C3	2	22 pf		
C4, C5	2	.01 µf		
Miscellaneous				
JP1—JP4	4	2-post male header connector—2.5-mm spacing		
JP5—motors	1	4-post male header connector—2.5-mm spacing		
JP6—RF module	1	4-post female header connector—2.5-mm spacing		

(continued on next page)

TABLE 6.2	Part	Quantity	Description
Parts List for Crocobot's Main Controller Board (continued)	Y1	1	4-MHz crystal
	W1-W4	4	Jumper wire
	Piezo buzzer	1	Standard piezoelectric element
	I.C. socket	1	18-pin I.C. socket—soldered to PC board U2
	Printed circuit board	1	See details in chapter.

L298 Dual Full-Bridge Driver

This robot is a departure from the previous two robots detailed in this book because it uses a twin DC motor gearbox as its source of power, instead of RC servos. In order to safely control the motors with the microcontroller, the L298 dual full-bridge driver will be used, and is shown in **Figure 6.33**. The L298 is an integrated monolithic circuit in a 15-lead multiwatt package. It is a high-voltage, high-current dual full-bridge driver designed to accept standard TTL logic levels and drive inductive loads such as relays, solenoids, DC, and stepping motors. Two enable inputs are provided to enable or disable the device independently of the input signals. The emitters of the lower transistors of each bridge are connected together, and the corresponding external terminal can be used for the connection of an external sensing resistor. An additional supply input is provided so that the logic functions at a lower voltage.

L298
Dual Full-Bridge Driver

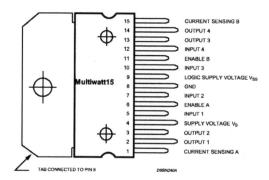

```
15    CURRENT SENSING B
14    OUTPUT 4
13    OUTPUT 3
12    INPUT 4
11    ENABLE B
10    INPUT 3
 9    LOGIC SUPPLY VOLTAGE Vss
 8    GND
 7    INPUT 2
 6    ENABLE A
 5    INPUT 1
 4    SUPPLY VOLTAGE Vs
 3    OUTPUT 2
 2    OUTPUT 1
 1    CURRENT SENSING A
```

Multiwatt15

TAB CONNECTED TO PIN 8   D95IN240A

FIGURE 6.33

L298 bidirectional motor controller.

How it works. The L298 contains two motor control circuits that are referred to as the "H-Bridge." This method of controlling DC motors gets its name because the four transistors used to control the motors are configured to form an "H" with the motor being at the center. **Figure 6.34** shows the basic schematic for a typical H-Bridge. The H-Bridge works by having the control circuitry or microcontroller turn on only two of the transistors at a time. In this example, when transistors Q1 and Q4 are turned on, the motor will spin in one direction. When transistors Q2 and Q3 are turned on, the motor will spin in the opposite direction. When all of the transistors are turned off, the motor is stopped. **Table 6.3** is a truth table showing the state of each transistor and the motor direction. Note that if transistors Q1 and Q3 (or Q2 and Q4) were turned on at the same time, there would be a short circuit across the battery. For this reason, the L298 has internal logic that prevents this from happening.

Motor direction	Q1	Q2	Q3	Q4
Stopped	0	0	0	0
Forward	1	0	0	1
Reverse	0	1	1	0

TABLE 6.3

H-Bridge Truth Table

FIGURE 6.34

A typical H-Bridge DC motor control configuration.

Motor power supply +

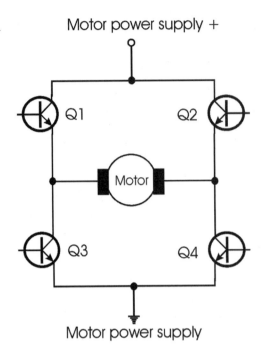

Motor power supply

With the L298, each bridge has three control inputs made up of an enable line and two control lines. In our robot application, these inputs will be controlled by the programmable interface controller (PIC). The PIC will interpret the data received by the radio link and then issue the proper motor commands, depending on the information sent from the hand remote control. An external bridge of diodes is required when inductive loads like DC motors are being driven. The specifics of controlling the motors will be described during the programming section.

Radio transmitter and receiver modules. The robot will be remotely controlled using a pair of 433-MHz transmitter and receiver modules. The modules that will be used are the TXLC-434 transmitter and the RXLC-434 receiver, available from Reynolds Electronics at www.rentron.com. The modules are based around Linx Technologies' (www.linxtechnologies.com) LC series transmitter modules. The staff at Reynolds Electronics have made using these devices very easy by mounting the modules on small circuit boards with connectors and a place to solder on the antennas (which are included with the modules).

The LC Series is ideally suited for volume use in applications such as remote control, security, identification, robotics, and periodic data transfer. Packaged in a compact SMD package, the LC transmitter utilizes a highly optimized SAW architecture to achieve an unmatched blend of performance, size, efficiency, and cost. When paired with a matching LC series receiver, a highly reliable wireless link is formed, capable of transferring serial data at distances in excess of 300 feet. No external RF components, except an antenna, are required, making design integration straightforward. The features include: low cost, no external RF components required, ultra-low power consumption, compact surface-mount package, stable SAW–based architecture, support data rates to 5,000 bps, wide supply range (2.7-5.2 vdc), direct serial interface, low harmonics, and no production tuning. The receiver module pinout diagram is shown in **Figure 6.35**. Using the module to receive information from the transmitter will be described when programming is covered.

FIGURE 6.35

Receiver module pinout diagram.

Creating the Main Controller Printed Circuit Board

To fabricate the controller printed circuit board (PCB), photocopy the artwork in **Figure 6.36** onto a transparency. Make sure that the photocopy is the exact size of the original. For convenience, you can download the file from the author's Web site, located at www.thinkbotics.com, and simply print the file onto a transparency using a laser or ink-jet printer with a minimum resolution of 600 dpi. After the artwork has been successfully transferred to a transparency, use the techniques outlined in Chapter 2 to create a board. A 4-inch × 6-inch presensitized positive copper board is ideal. When you place the transparency on the copper board, it should be oriented exactly the same as in **Figure 6.36**. It would be a good idea to create the circuit board for the remote control at the same time.

FIGURE 6.36

Controller board PCB foil pattern artwork.

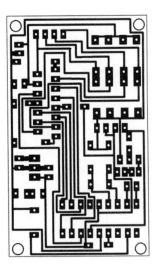

Circuit board drilling and parts placement. Use a 1/32-inch drill bit to drill all of the component holes on the PCB. Drill the holes for the voltage regulator (U1) and the diodes (D2–D9) with a 3/64-inch drill bit. Use **Table 6.2** and **Figure 6.37** to place the parts on the component side of the circuit board. The PIC 16F84 microcontroller (U2) is mounted in an 18-pin I.C. socket. The 18-pin socket is soldered to the PC board, and the PIC is inserted after it has been programmed. Note that **Figure 6.37** also shows four jumper wires labeled W1–W4 that are not shown in the schematic. These jumpers were needed due to routing conflicts when designing the PCB. Use a fine-toothed saw to cut the board along the guide lines, and drill the mounting holes on the corners using a 5/32-inch drill bit. Use 1/4-inch standoffs to mount the board. **Figure 6.38** shows the finished main controller board.

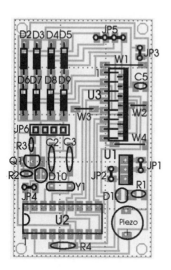

FIGURE 6.37

Controller board PCB component side parts placement.

FIGURE 6.38

Parts soldered to the finished PCB.

Check the finished board for any missed or cold soldered connections, and verify that all the components have been included. The board will be tested later when programming the PIC microcontroller.

Adding the radio receiver module. Locate the radio receiver module (RXLC-434) and flip it over so that the back is facing upward. Solder the 7-inch antenna wire that was included with the module to the tinned area on the board where there is no solder mask. **Figure 6.39** shows the antenna soldered to the board.

The next step is to bend all of the connector pins of the receiver module on 90-degree angles toward the back of the module. Use a pair of needle nose pliers to carefully bend each pin. This is needed so that the module will sit parallel to the controller board when it is plugged into its connector. **Figure 6.40** illustrates how

Solder antenna here

FIGURE 6.39

Antenna soldered to the receiver module PCB.

FIGURE 6.40

Receiver module connector pins bent 90 degrees.

the pins should be bent. Once the pins have been bent, insert the module into the 4-pin female connector (JP6) located in front of the diode array. Orient the module so that it sits above the diodes when it is plugged in. **Figure 6.41** show the module plugged into the circuit board.

FIGURE 6.41

Receiver module
inserted into connector
on the main board.

Putting It All Together

Now that the mechanical, electronics, and electrical systems are all finished, it is time to integrate them all together into a working robot. Start by mounting the circuit board to the chassis at the head of the robot. Attach the robot's tail section to the chassis with a 6/32-inch × 1/2-inch machine screw and locking nut. Tighten the nut with enough torque to let the tail swing freely. Plug each of the connectors into the main controller, as indicated in **Figure 6.42**. Note that the motor power supply battery pack can't be connected until the top cover has been attached to the chassis. Place a new AA battery into each of the three battery holders located on the top cover. **Figure 6.43** shows the robot with the tail section attached and all of the connecting wires plugged into the controller board. Place a 9-volt battery into the battery clip located in the tail section. Attach the battery strap to the battery. Feed the antenna through the hole in the head section, then use three 6/32-inch × 1/2-inch machine screws and nuts to attach the top cover. Plug in the motor power connector before you fasten the cover in place. **Figure 6.44** shows the completed robot with the

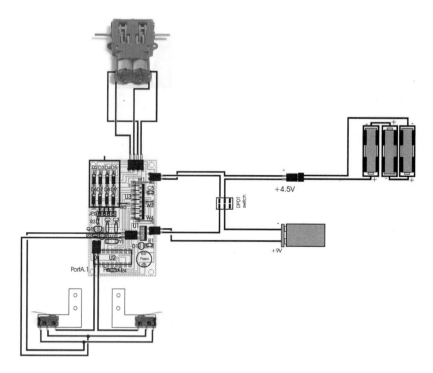

FIGURE 6.42

Robot connection diagram.

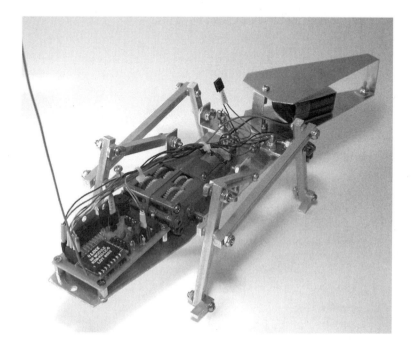

FIGURE 6.43

Robot with tail section attached and all wiring connected.

FIGURE 6.44

Completed robot with cover attached.

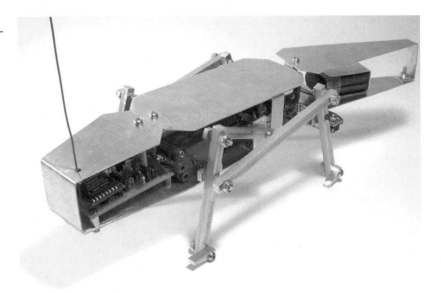

top cover attached. The PIC microcontroller will be programmed a little later, during experimentation. Now that the robot is complete, the remote control transmitter will be built.

Constructing the Remote Control Transmitter

The remote control transmitter will be used to control the robot's movements and may be customized to control other devices as well. The hand held remote control device uses an analog X and Y axis control stick as the input to two analog-to-digital converters residing on a PIC 16C71. The remote control is pictured in **Figure 6.45**.

FIGURE 6.45

Robot remote control device.

The schematic for the transmitter remote control is shown in **Figure 6.46**. The circuit functions by using the PIC 16C71 to monitor the position of the control stick and then send serial commands to the transmitter module. When the control stick moves along the X and Y axis, the resistance values of two 100K Ω potentiometers are varied. The control stick and the two attached potentiometers are shown in **Figure 6.47**. Each potentiometer is configured as a voltage divider so that a unique voltage represents each position along the X- and Y-axis. The voltages from the potentiometers are converted to 8-bit values by the internal analog to digital converters on the PIC 16C71 and then interpreted by the microcontroller. Depending on the values, certain movement commands are sent in a serial format from the transmitter to the robot. The remote control also has a programmable push-button switch and a light-emitting diode (LED) that can be turned on when certain events occur, such as during the transmission of a movement command. The transmit-

FIGURE 6.46

Remote control
schematic diagram.

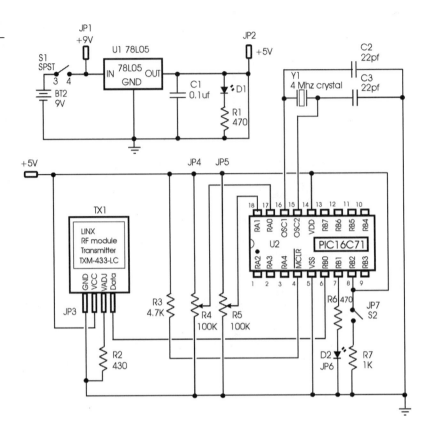

ter module is the TXLC-434 transmitter, available from Reynolds Electronics at: www.rentron.com. The modules are based around Linx Technologies' (www.linxtechnologies.com) LC series transmitter modules, as discussed earlier. The transmitter module pinout diagram is shown in **Figure 6.48**. The only external part needed for the module to function is a 430 Ω resistor that is connected from the VADJ line to ground for 5-volt operation. If the resistor is not included, then the device will operate at 3 volts. Using the module to transmit information to the receiver will be discussed when programming is covered.

FIGURE 6.47

Control stick with X and Y axis potentiometers.

FIGURE 6.48

Transmitter module pinout diagram.

PIC 16C71

The Microchip PIC 16C71 is very similar to the PIC 16F84 that has been used throughout the book. The pinouts are identical. The difference is that the pins on PortA of the 16C71 can be configured to take advantage of four on-chip analog-to-digital converters. Another difference is that the chip is erased by exposure to ultraviolet light. A small window on the top of the device allows light to get at the chip. After the chip has been programmed, the window should be covered with a sticker so that it does not get erased if it is exposed to sunlight or fluorescent lighting. The 8-bit resolution of the 4-channel high-speed 8-bit A/D is ideally suited for applications requiring a low-cost analog interface. Use of the A/D converters will be discussed when the software routines are covered. Although the 16C71 device was used in the book, Microchip now manufactures an 18-pin, flash erasable device with analog-to-digital converters, identified as the PIC 16F818. **Figure 6.49** shows the PIC 16C71 with its ultraviolet erase window. The parts needed to build the transmitter are listed in **Table 6.4**.

FIGURE 6.49

Microchip PIC 16C71.

Part	Quantity	Description	**TABLE 6.4**
			List of Parts Needed to Build the Transmitter
Semiconductors			
U1	1	78L05 5V regulator	
U2	1	PIC 16C71 microcontroller mounted in socket	
TX1	1	Lynx TXM-433-LC-R RF transmitter module	
D1	1	Red light-emitting diode	
D2	1	Red light-emitting diode	
Resistors			
R1,R2,R6	3	470 Ω 1/4-watt resistor	
R3	1	4.7 KΩ 1/4-watt resistor	
R4,R5	2	Control stick with two 100 KΩ potentiometers	
R7	1	1 KΩ 1/4-watt resistor	
Capacitors			
C1	1	0.1 µf	
C2,C3	2	22 pf	
Miscellaneous			
JP1	1	2-post male header connector—2.5-mm spacing	
JP2,JP6,JP7	3	2-post female header connector—2.5-mm spacing	
JP3	1	4-post female header connector—2.5-mm spacing	
JP4,JP5	2	3-post female header connector—2.5-mm spacing	

(continued on next page)

TABLE 6.4	Part	Quantity	Description
List of Parts Needed to Build the Transmitter (continued)	Y1	1	4-MHz crystal
	I.C. socket	1	18-pin I.C. socket—soldered to PC board U2
	Project box	1	3 inches wide x 1-1/2 inches deep
	Battery strap	1	9-volt battery strap
	S1—switch	1	SPST switch
	S2—switch	1	Momentary contact—normally open pushbutton
	Antenna	1	6-3/4 inch whip antenna with threaded mount
	Enclosure connectors		
	JP1	1	2-post female header connector—2.5-mm spacing
	JP2,JP6,JP7	3	2-post male header connector—2.5-mm spacing
	JP3	1	4-post male header connector—2.5-mm spacing
	JP4,JP5	2	3-post male header connector—2.5-mm spacing

Creating the Remote Control Printed Circuit Board

To fabricate the PCB, photocopy the artwork in **Figure 6.50** onto a transparency. Make sure that the photocopy is the exact size of the original. For convenience, you can download the file from the author's Web site, located at www.thinkbotics.com, and simply print the file onto a transparency using a laser or ink-jet printer with a minimum resolution of 600 dpi. After the artwork has been

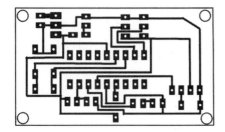

FIGURE 6.50

Remote control PCB foil pattern artwork.

successfully transferred to a transparency, use the techniques outlined in Chapter 2 to create a board. A 4-inch × 6-inch presensitized positive copper board is ideal. When you place the transparency on the copper board, it should be oriented so that it is exactly the same as in **Figure 6.50**.

Circuit board drilling and parts placement. Use a 1/32-inch drill bit to drill all of the component holes on the PCB. Drill the holes for the voltage regulator (U1) with a 3/64-inch drill bit. Use **Table 6.4** and **Figure 6.51** to place the parts on the component side of the circuit board. Note that female sockets are used where certain components will be plugged in. This is to make it easier to mount the control potentiometers, LEDs, and switches to the top cover of the project box. The PIC 16C71 microcontroller (U2) is mounted in an 18-pin I.C. socket. The 18-pin socket is soldered to the PC board, and the PIC is inserted after it has been programmed. Use a fine-toothed saw to cut the board along the guide lines.

Check the finished board for any missed or cold soldered connections, and verify that all the components have been included. The board will be tested later when programming the PIC microcontroller.

FIGURE 6.51

Remote control PCB
component side parts
placement.

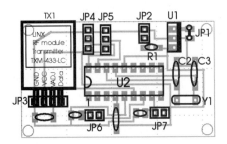

Remote control project enclosure. Choose a project box that is at least 3 inches wide, 5 inches in length, and 1-1/2 inches deep. Depending on the control stick that you are using, the box may need to be larger or smaller than the dimensions above. I used a project box that had removable top and bottom panels to make it easier to work with.

Locate the 6-3/4 inch whip antenna and cut the coaxial cable to a length of 2-1/2 inches in length. Strip 1/2-inch of the shielding off the end of the wire, and then strip the middle wire as well. Drill a 1/4-inch hole in the top, right side of the case, and mount the antenna. Solder the antenna lead wire to the small area on the back (the area without any solder mask) of the transmitter module. Bend the connector pins of the transmitter module 90 degrees downward. This is the same procedure that was performed on the receiver module. Place the remote control circuit board in the case, and then plug the transmitter module into the female connector (JP3). Move the circuit to the top of the case, 1/2-inch from the top. Use hot glue to secure the board in place. **Figure 6.52** shows the finished transmitter circuit board, with the antenna attached to the case and the transmitter module.

FIGURE 6.52

Remote control circuit board with antenna connected.

Mount the control stick, power switch, two LEDs, and push-button switch to the top cover of the project box in similar positions, as shown in **Figure 6.53**. You will have to drill a 3/4-inch hole for the control stick. Depending on the project box that you are using, you may have to find the best positions for each of the components. When the parts are mounted in the cover, use **Figure 6.54** to wire the parts to the board. I used wires with a length of 3-1/2 inches to connect each component to the appropriate connector. **Figure 6.55** shows the components wired to the connectors. Once the parts are wired to the connectors, attach a 9-volt battery, but move the cover to the side to leave access to the 18-pin socket, so that the PIC 16C7l can easily be inserted and removed during the programming, debugging, and experimentation stages. We are now ready to start programming the robot and transmitter.

FIGURE 6.53

Mounting placement of control stick, switches, and light emitting diodes.

FIGURE 6.54

Transmitter wiring diagram.

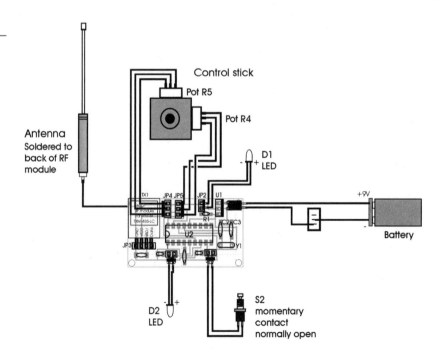

FIGURE 6.55

Transmitter components wired to connectors.

Programming Crocobot

To bring the crocodile robot to life, the leg sensor switches will be checked to make sure that they are working properly. The leg sensor switches will be used to coordinate the walking gait of the robot. With the two-motor, four-leg design that has been used, it is necessary for one set of legs to be in the forward position when the other set of legs are in motion. Otherwise, the robot does not get maximum body lift or forward/reverse motion. The first program that will be written is called crocobot-switch.bas and is listed in **Program 6.1**. Enter the program into your favorite text editor, then compile and program the PIC 16F84 using the crocobot-switch.hex file listed in **Program 6.2**. Insert the PIC into the 18-pin socket on the main board. Move the legs by hand so that the limit switches are not triggered, and then turn on the power. The robot should make a start-up sound and then go silent. If tones are being produced without the switch being pushed, then turn the power off. Make sure that the polarity of JP2 is correct, with +5 being connected to the middle connector of the limit switches (see **Figure 6.42**). Turn the power back on, and push the limit switch on the robots right side with your finger. The PIC should produce alternating high and low tones from the piezo speaker. Trigger the left switch with your finger. A steady pulsing tone should be heard. If the tones being produced do not correspond to the correct switch, then reverse connector JP4 (**Figure 6.42**).

PROGRAM 6.1

crocobot-switch.bas program listing

```
'_____
' Name       : croco-switch.bas
' Compiler   : PicBasic Pro - MicroEngineering Labs
' Notes      : Program to test the leg limit switches
'_____


' PortA set as outputs. pins 0 and 1 inputs
trisa = %00000011

' PortB set as outputs. pin 1 input.
trisb = %00000001


'_____

' initialize variables

include "modedefs.bas"

limit_left      VAR PORTA.0
limit_right     VAR PORTA.1
piezo           VAR PORTA.3

SOUND PIEZO,[115,10,50,10]

start:

If limit_left = 1 then
   SOUND PIEZO,[100,10]
   pause 20
endif
```

If limit_right = 1 then
   SOUND PIEZO,[80,20]
   pause 40
   SOUND PIEZO,[110,20]
   pause 40
endif

goto start

end

PROGRAM 6.1

crocobot-switch.bas
program listing
(continued)

```
:1000000061288F0022088400200928208413 8F088B
:1000100003195C28F03091000E0880389000F03011
:10002000910303199100031 98F0303195C28182801
:100030002B2003010C1820088E1F20088E0803199E
:10004000301900F252880060C28262800000F2881
:1000500084178 0055C280D080C0403198C0A803075
:100060000C1A8D060C198D068C188D060D0D8C0D35
:100070008D0D5C288F018E00FF308E07031C8F07CB
:10008000031C5C2803308D00DF3048203C288D01A4
:10009000E83E8C008D09FC30031C51288C070318A6
:1000A0004E288C0764008D0F4E280C1857288C1C86
:1000B0005B2800005B2808008313031383126 4008D
:1000C000080083160330850001308600053083125 6
:1000D000A2000830A00073308E000A3001203230B8
:1000E0008E000A30012064000 51C80280530A20023
:1000F0000830A00064308E000A30012014303A200D
:1001000006400851C97280530A2000830A0005030FC
:100110008E001430012028303A200530A20008302B
:10012000A0006E308E001430012028303A20732851
:0401300063009828A8
:02400E00F53F7C
:00000001FF
```

PROGRAM 6.2

crocobot-switch.hex file
listing

The next program will test the robot leg motors and will ensure that the motor connector is oriented correctly. Compile motor-test.bas, listed in **Program 6.3**. Program the PIC 16F84 with the motor-test.hex file listed in **Program 6.4**, and then insert it into the 18-pin socket on the main board. When the power is turned on, the left leg motor should rotate in a forward direction for 3 seconds, and then turn off. The right leg motor should then rotate forward for 3 seconds. The whole sequence will then repeat itself. If either the left or right leg motors are rotating in the reverse direction during this test, then de-solder the wires connected to the offending motor, reverse them, and re-solder the connections. If the first motor to run is the right motor, then unplug the motor connector (JP3), turn it around, and plug it back in. The motors are controlled by first setting the L298 enable pins high on the desired channels. The forward or reverse pins for each channel are then set high, depending on the direction in which you want the motor to travel. If both the forward and reverse pins are set high, then the chip will perform a fast motor stop.

PROGRAM 6.3

motor-test.bas program listing

```
'_____
'  Name     : motor-test.bas
'  Compiler : PicBasic Pro - MicroEngineering Labs
'  Notes    : Program to test the leg motors
'_____


' PortA set as outputs. pins 0 and 1 inputs
trisa = %00000011

' PortB set as outputs. pin 1 input.
trisb = %00000001


'_____

' initialize variables

include "modedefs.bas"
```

PROGRAM 6.3

motor-test.bas program
listing (continued)

```
enable_right    VAR PORTB.1
forward_right   VAR PORTB.2
reverse_right   VAR PORTB.3
enable_left     VAR PORTB.4
reverse_left    VAR PORTB.5
forward_left    VAR PORTB.6
limit_left       VAR PORTA.0
limit_right      VAR PORTA.1
piezo            VAR PORTA.3

low enable_left
low forward_left
low reverse_left

low enable_right
low forward_right
low reverse_right


SOUND PIEZO,[115,10,50,10]

start:

high enable_left
high forward_left
pause 3000
low enable_left
low forward_left

high enable_right
high forward_right
pause 3000
low enable_right
low forward_right

goto start

end
```

PROGRAM 6.4

motor-test.hex file
listing

```
:1000000061288F0022088400200928208413 8F088B
:100010003195C28F03091000E0880389000F03011
:10002000910303199100031 98F0303195C28182801
:100030002B2003010C1820088E1F20088E0803199E
:10004000030190 0F252880060C28262800000F2881
:1000500084178005 5C280D080C0403198C0A803075
:100060000C1A8D060C198D068C188D060D0D8C0D35
:100070008D0D5C288F018E00FF308E07031C8F07CB
:10008000031C5C2803308D00DF3048203C288D01A4
:10009000E83E8C008D09FC30031C51288C070318A6
:1000A0004E288C0764008D0F4E280C1857288C1C86
:1000B0005B2800005B2808008313031383126400 8D
:1000C00008008316033085000130860083120612 73
:1000D0000831606128312061383160613831286 12E2
:1000E0000831686128312861083168610831206 11D9
:1000F0000831606118312861183168611053083122A
:10010000A2000830A00073308E000A3001 20323087
:100110008E000A3001200616831606128312061777
:100120000831606138312 0B308F00B8303B20061263
:1001300008316061283120613831606138312861 47F
:1001400008316861083120615831606118312 0B3050
:100150008F00B8303B208610831686108312 06115C
:0C0160000831606118312 8B286300B4285C
:02400E00F53F7C
:00000001FF
```

In the next program, the robot's four basic walking subroutines
will be developed. The robot will walk forward, turn to the left,
walk in reverse, and then turn to the right. As described earlier,
in order for a two-motor, four-legged robot to walk successfully,
it is necessary for one set of legs to be in the forward position
when the other set of legs are in motion. This ensures static sta-
bility. Otherwise, the robot does not get maximum body lift or
forward/reverse motion with each leg cycle. For example, when
the crocodile robot moves in a forward direction, the microcontroller first turns on the left leg motor. A small delay is introduced so that the leg has time to move past the leg position

switch, since that was possibly the position that is was stopped at during the last cycle. The program goes into a tight while loop to monitor the leg switch. When the leg makes a complete cycle, the leg switch is activated, program execution breaks out of the while loop, and the leg motor is turned off. The microcontroller then goes through the same logic with the right leg moving in a forward direction. With both the left leg and the right leg moving for one cycle in the forward direction, the robot's body is moved forward. This logic works for most cases, but depending on the position of the leg during the last leg cycle, the leg may actually trigger the switch right away, and a full leg cycle does not occur. This is taken care of by having each routine run twice, so that if a cycle was missed on the first time through, it will occur the second time through.

Each of the walking routines uses the logic stated above. In order for the robot to walk in reverse, both legs move in the reverse direction. To turn the robot to the right, the left leg moves forward and the right leg moves in reverse. To turn the robot to the left, the left leg moves in reverse, while the right leg moves forward. Any of the four walking routines can be combined to diversify the movement. An example of this would be if you wanted the robot to move forward and to the right. This would be accomplished by calling the forward subroutine and then the right subroutine, alternating between the two.

Program 6.5 is called walk-routines.bas. This program demonstrates each of the four walking routines that will be used later with the remote control. Program the PIC 16F84 with the walk-routines.hex file listed in **Program 6.6**. When the program executes, the robot will run through the forward routine five times, for a total of 10 leg cycles. It will then turn to the left, walk in reverse, and then turn to the right. Now that the walking routines have been developed, the radio remote control can be added.

PROGRAM 6.5

walk-routines.bas
program listing

```
'_____

' Name     : walk-routines.bas
' Compiler : PicBasic Pro - MicroEngineering Labs
' Notes    : various walking subroutines
'_____


' PortA set as outputs. pins 0 and 1 inputs
trisa = %00000011

' PortB set as outputs. pin 1 input.
trisb = %00000001


'_____

' initialize variables

include "modedefs.bas"

enable_right    VAR PORTB.1
forward_right   VAR PORTB.2
reverse_right   VAR PORTB.3
enable_left     VAR PORTB.4
reverse_left    VAR PORTB.5
forward_left    VAR PORTB.6
limit_left       VAR PORTA.0
limit_right      VAR PORTA.1
piezo            VAR PORTA.3
temp             VAR BYTE

low enable_left
low forward_left
low reverse_left

low enable_right
low forward_right
low reverse_right

SOUND PIEZO,[115,10,50,10]

start:
```

PROGRAM 6.5

walk-routines.bas
program listing
(continued)

```
'_____

' walking subroutines

walk_forward:

    For temp = 1 to 5

    ' move left leg

    high enable_left
    high forward_left
    pause 300

    while limit_left = 0
    wend

    low enable_left
    low forward_left

    ' move right leg

    high enable_right
    high forward_right
    pause 300

    while limit_right = 0
    wend

    low enable_right
    low forward_right

    next temp


'_____

turn_left:

    For temp = 1 to 5
```

PROGRAM 6.5

walk-routines.bas
program listing
(continued)

```
' move left leg

high enable_left
high reverse_left
pause 300

while limit_left = 0
wend

low enable_left
low reverse_left

' move right leg

high enable_right
high forward_right
pause 300

while limit_right = 0
wend

low enable_right
low forward_right

next temp

'_____

walk_reverse:

    For temp = 1 to 5

    ' move left leg

    high enable_left
    high reverse_left
    pause 300
```

PROGRAM 6.5

walk-routines.bas
program listing
(continued)

```
while limit_left = 0
wend

low enable_left
low reverse_left

' move right leg

high enable_right
high reverse_right
pause 300

while limit_right = 0
wend

low enable_right
low reverse_right

next temp


'_____


turn_right:

For temp = 1 to 5

    ' move left leg

high enable_left
high forward_left
pause 300

while limit_left = 0
wend

low enable_left
low forward_left
```

PROGRAM 6.5

walk-routines.bas
program listing
(continued)

```
' move right leg

high enable_right
high reverse_right
pause 300

while limit_right = 0
wend

low enable_right
low reverse_right

next temp

goto start

end
```

PROGRAM 6.6

walk-routines.hex file
listing

```
:1000000061288F00220884002009282084138F088B
:1000100003195C28F03091000E0880389000F03011
:100020009103031991000319 8F0303195C28182801
:100030002B2003010C1820088E1F20088E0803199E
:100040000301900F252880060C28262800000F2881
:1000500084178005 5C280D080C0403198C0A803075
:100060000C1A8D060C198D068C188D060D0D8C0D35
:100070008D0D5C288F018E00FF308E07031C8F07CB
:10008000031C5C2803308D00DF3048203C288D01A4
:10009000E83E8C008D09FC30031C51288C070318A6
:1000A0004E288C0764008D0F4E280C1857288C1C86
:1000B0005B2800005B280800831303138312640 08D
:1000C000080083160330850001308600831206 1273
:1000D0008316061283120613831606138312861 2E2
:1000E00083168612831286108316861083120611D9
:1000F00083160611831286118316861105308312 2A
:10010000A2000830A0007330 8E000A30012032308 7
:100110008E000A3001200130A4006400063024026 1
:10012000318C4280616831606128312061783 16B0
```

```
:10013000061383120130 8F002C303B206400051819
:10014000A2289E280612831606128312061383160F
:100150000613831286148316861083120615 8316DF
:10016000061183120130 8F002C303B20640008518 6B
:10017000BA28B6288610831686108312 06118316B5
:1001800006118312A40F8D280130A40064000630EC
:1001900024020318FD280616831606128312 8616FB
:1001A000831686128312 0130 8F002C303B206400AE
:1001B0000518DB28D7280612831606128312 86122A
:1001C0008316861283128614831686108312 0615F0
:1001D00008316061183120130 8F002C303B206400FF
:1001E0008518F328EF2886108316861083120611CF
:1001F000083160611 8312A40FC6280130A4006400 0E0
:100200000063024020318362906168316061283 12B6
:10021000086168316861283120130 8F002C303B2005
:100220000064000518142910290612831606128312 79
:1002300008612831686128312861483168610831202
:10024000086158316861183120130 8F002C303B20D7
:100250000640085182C29282986108316861083129D
:10026000086118316861183 12A40FFF280130A40083
:10027000064000630240203186F290616831606123E
:100280000831206178316061383120130 8F002C3059
:10029000 03B2064000518 4D294929 0612831606 12D1
:1002A0000083120613831606138312861483168610 90
:1002B0000831286158316861183120130 8F002C302D
:1002C000 03B20640085186529612986108316861 0F5
:1002D0000831286118316861183 12A40F38298B2866
:0402E000063007 0291E
:02400E00F53F7C
:00000001FF
```

PROGRAM 6.6

walk-routines.hex file
listing (continued)

The Lynx LC series transmitter and receiver modules were designed to facilitate a highly reliable wireless serial data link. Once the units are powered up, serial data can be sent on a single pin of the transmitter and received on a single pin of the receiver. This makes it very easy to use the modules with a microcontroller and a programming language like PicBasic Pro.

When using the serin command to receive data, PicBasic Pro lets you define a qualifier enclosed within brackets before any more data is received. Serin must receive these bytes in exact order before receiving any data items. If any bytes received do not match the next byte in the qualifier sequence, the qualification process starts over (i.e., the next received byte is compared to the first item in the qualifier list). This makes it easy for us to program an identification code for each device or robot that we are going to control. It also ensures that good data are being sent, and cuts down on erroneous interpretation of the received serial data. Once the qualifiers are satisfied, serin begins storing data in the variables associated with each item. If the variable name is used alone, the value of the received ASCII character is stored in the variable.

For our test program, we will use the qualifier "Z" to identify that the received data is coming from our transmitter. Once the character "Z" has been received, the next byte of information will be stored in a variable. We can then compare this information to certain control commands and carry out the tasks associated to them. The line of code that will receive the serial data looks like this:

```
serin rxmit,rx_baud,["Z"],control
```

The serial data is received on pin rxmit (PORTB.0) at a baud rate of rx_baud (2400). The program execution will remain in a tight loop, receiving serial data and comparing each received byte to the character "Z." When the character "Z" is received, the next data byte is stored in the variable named control. We will then compare the contents of the variable control to the character "A." If the comparison is true, then the microcontroller will produce a couple of tones on the piezo speaker. The receiver program is called receive-test.bas and is listed in **Program 6.7**. Program the PIC 16F84 with the receive-test.hex file listed in **Program 6.8**. Place the PIC in the 18-pin socket on the robot's main board.

PROGRAM 6.7

receive-test.bas
program listing

```
'_____
' Name     : receive-test.bas
' Compiler : PicBasic Pro - MicroEngineering Labs
' Notes    : Program to test the wireless data link
'            : between the Lynx 433LC series
'            : transmitter and receiver.
'_____


' PortA set as outputs. pins 0 and 1 inputs
trisa = %00000011

' PortB set as outputs. pin 0 input.
trisb = %00000001


'_____

' initialize variables

include "modedefs.bas"

rx_baud      CON N2400
rxmit        VAR PORTB.0
piezo        VAR PORTA.3
control      VAR BYTE

SOUND PIEZO,[115,10,50,10]

start:

serin rxmit,rx_baud,["Z"],control

if control = "A" then
   SOUND PIEZO,[115,10,80,20]
   pause 100
endif

goto start

end
```

PROGRAM 6.8

receive-test.hex file
listing

```
:100000009F2864001120031801281C2008308F004D
:100010001D2011208E0C0C288F0B08281D200E0887
:1000200008002208840020088417800484130005 37
:100030001F192006FF3E08001F171F0D06398C00F0
:100040002B208D008C0A2B201F1F7E281F138C0055
:1000500002309 5207E2800308A000C088207013487
:1000600075340334153400343C340C34D9348F00E7
:10007000220884002009 5E2084138F0803199A281F
:10008000F03091000E0880389000F0309103031991
:100090009100031 98F0303199A284E286120030148
:1000A0000C1820088E1F20088E0803190301900FDA
:1000B0005B28800642285C280000452884178005BC
:1000C0009A280D080C0403198C0A80300C1A8D062E
:1000D0000C198D068C188D060D0D8C0D8D0D9A2822
:1000E0008F018E00FF308E07031C8F07031C9A2898
:1000F00003308D00DF307E2072288D01E83E8C00B9
:100100008D09FC30031C87288C07031884288C0772
:1001100064008D0F84280C188D288C1C91280000F9
:100120009128080003108D0C8C0CFF3E03189228B8
:100130000C089A28831303138312640008008316A3
:1001400003308500013086000530831 2A20008309C
:10015000A00073308E000A30372032308E000A3013
:100160003720063 0A2000130A00004309F0001209B
:100170005A3C031DB7280120A40064002408413C18
:10018000031DD0280530A2000830A00073308E0077
:100190000A30372050308E00143037206430702001
:0601A000B1286300D12824
:02400E00F53F7C
:00000001FF
```

The corresponding transmit-test.bas for the PIC 16C71 microcontroller used in the remote control is listed in **Program 6.9**. This program uses the serout command to send serial data to the transmitter. The baud rate is also set at the same rate as the receiver program. Notice that the qualifier character "Z" is sent first, and then our control character, in this case "A." Program the PIC 16C71 with the transmit-test.hex file listed in **Program 6.10**.

Insert the 16C71 into the 18-pin socket on the remote control circuit board. Turn the power on at the robot, and then turn on the power to the remote control. When the button on the remote control is pushed, the LED above the button will light up, indicating that a transmission has been sent. At the same time, the piezo speaker on the robot will make a couple of tones each time the button is pushed. You should also notice that the LED next to the receiver module on the robot's controller board will flash on and off rapidly, as data comes through. If nothing happens when the button is pushed, check all of your wiring and battery supplies. Once the units are working together correctly, you can check the range of the transmitter by walking away from the robot and holding the push button on the remote control down. For later experimentation, you can program this button for other tasks.

PROGRAM 6.9

transmit-test.bas
program listing

```
' _____

' Name      : transmit-test.bas
' Compiler  : PicBasic Pro - MicroEngineering Labs
' Notes     : Program to test wireless link using the
'           : Linx 433LC series transmitter and receiver
' _____

' set PortA  inputs
trisa = %00011111

' PortB set as outputs. pin 2 input
trisb = %00000100


' _____

' initialize variables

include "modedefs.bas"

tx_baud        CON N2400

txmit          VAR PORTB.0
```

PROGRAM 6.9

transmit-test.bas
program listing
(continued)

```
txmit_led       VAR PORTB.1
push_button     VAR PORTB.2

start:

low txmit_led

If push_button = 1 then
    serout txmit,tx_baud,["ZA"]
    high txmit_led
    pause 200
endif

goto start
```

PROGRAM 6.10

transmit-test.hex file
listing

```
:10000000632892002208840009309300310000D2019
:10001000920C930B072803140D2884139F1D1C2892
:100020000008200041F1D2006800084170008200 4FB
:1000300031C200680002728000820040031C20063B
:100040001F192006800084172009800527281F0D0E
:100050006398C0030208D008C0A302000004A28A0
:1000600000308A000C08820701347534033415 34DB
:1000700000343C340C34D9348F018E00FF308E07AD
:1000800031C8F07031C5E2803308D00DF304A20DD
:100090003E288D01E83E8C008D09FC30031C53285E
:1000A0008C07031850288C0764008D0F50280C18FB
:1000B00059288C1C5D2800005D2808008313031359
:1000C00083126400080083161F3085000430860008
:1000D00083128610831686106400831206 1D802802
:1000E000630A2000130A00004309F005A300120E9
:1000F000413001208614831686100C83083123C20BC
:0201000069286C
:02400E00F53F7C
:00000001FF
```

At this stage, we can bring all of the subroutines together into one set of robot remote control programs. The only thing left to discuss is the use of the analog-to-digital (A/D) converters on the PIC

16C71. These A/D converters will be used to convert the voltages from the control stick potentiometers to 8-bit digital values. Each potentiometer is configured as a voltage divider so that a unique voltage represents each position along the X and Y axis. The PicBasic Compiler also makes using the A/D converters very easy. Using the ADCIN command, it is easy to set the number of bits in the result, set the clock source, set the sampling rate, and set the port pins to analog. Once that has all been set up, simply read the channel value and store the result in a variable. I have listed all of the A/D converter registers in the comments of the transmitter code if you are interested in exactly what is happening.

The program for the robot is called rx-remote.bas and is listed in **Program 6.11**. Compile the code and then program the PIC 16F84 with the rx-remote.hex file listed in **Program 6.12**. Insert the programmed 16F84 into the 18-pin socket on the robot's main board. The program for the remote control is called tx-remote.bas and is listed in **Program 6.13**. Make sure that the PIC 16C71 has been U.V. erased. Compile the code and then program the PIC 16C71 with the tx-remote.hex file listed in **Program 6.14**. Insert the programmed 16C71 into the 18-pin socket on the remote control circuit board. Place the robot on the floor and turn on the power. Turn on the power to the remote control. Push the button on the front of the remote. The robot should make a sound. Try controlling the robot's direction using the control stick. When everything is working correctly, place the top on the transmitter project enclosure and secure it in place with the screws that came with the box.

With the control stick sitting in the middle position, the robot will be stopped. With the stick pushed all the way forward, the robot will walk forward. When the control stick is pulled backwards, the robot will walk in reverse. When the control stick is positioned to the right, the robot will turn to the right, and when the stick is positioned to the left, the robot will turn to the left. The potentiometer values were determined by taking the A/D readings and then outputting the values to an LCD display. You can check the program listing for the values. Feel free to make any changes or improvements. By using a serial wireless data link, the options are unlimited, so have fun with it.

PROGRAM 6.11

rx-remote.bas program listing

```
'_____

' Name      : rx-remote.bas
' Compiler  : PicBasic Pro - MicroEngineering Labs
' Notes     : Robot remote control using the Linx
'           : 433LC series transmitter and receiver.
'_____


' PortA set as outputs
trisa = %00000000

' PortB set as outputs. pin 0 input.
trisb = %00000001


'_____


' initialize variables

include "modedefs.bas"

rx_baud         CON N2400
rxmit               VAR PORTB.0
enable_right    VAR PORTB.1
forward_right   VAR PORTB.2
reverse_right   VAR PORTB.3
enable_left     VAR PORTB.4
reverse_left    VAR PORTB.5
forward_left    VAR PORTB.6
limit_left          VAR PORTA.0
limit_right         VAR PORTA.1
piezo               VAR PORTA.3
control             VAR BYTE
temp                VAR BYTE
```

```
low enable_left
low forward_left
low reverse_left

low enable_right
low forward_right
low reverse_right

SOUND PIEZO,[115,10,50,10]

start:

serin rxmit,rx_baud,["Z"],control

if control = "A" then
   gosub walk_forward
endif

if control = "B" then
   gosub walk_reverse
endif

if control = "C" then
   gosub turn_left
endif

if control = "D" then
   gosub turn_right
endif

if control = "E" then
   sound piezo,[115,10,50,10]
endif

if control = "F" then
   low enable_left
   low forward_left
   low reverse_left
```

PROGRAM 6.11

rx-remote.bas program
listing (continued)

PROGRAM 6.11

rx-remote.bas program
listing (continued)

```
        low enable_right
        low forward_right
        low reverse_right
    endif

    goto start


    '_____

    ' walking subroutines

    walk_forward:

        ' move left leg

        high enable_left
        high forward_left
        pause 300

        while limit_left = 0
        wend

        low enable_left
        low forward_left

        ' move right leg

        high enable_right
        high forward_right
        pause 300

        while limit_right = 0
        wend

        low enable_right
        low forward_right

    return
```

```
'_____

turn_left:

    ' move left leg

    high enable_left
    high reverse_left
    pause 300

    while limit_left = 0
    wend

    low enable_left
    low reverse_left

    ' move right leg

    high enable_right
    high forward_right
    pause 300

    while limit_right = 0
    wend

    low enable_right
    low forward_right

return

'_____


walk_reverse:

    ' move left leg
```

PROGRAM 6.11

rx-remote.bas program
listing (continued)

```
high enable_left
high reverse_left
pause 300

while limit_left = 0
wend

low enable_left
low reverse_left

' move right leg

high enable_right
high reverse_right
pause 300

while limit_right = 0
wend

low enable_right
low reverse_right

return

'_____


turn_right:

    ' move left leg

high enable_left
high forward_left
pause 300

while limit_left = 0
wend

low enable_left
```

```
low forward_left

' move right leg

high enable_right
high reverse_right
pause 300

while limit_right = 0
wend

low enable_right
low reverse_right

return

end
```

PROGRAM 6.11

rx-remote.bas program
listing (continued)

```
:100000009F286400112003180128 1C2008308F004D
:100010001D2011208E0C0C288F0B08281D200E0887
:1000200008002208840020088417 80048413000537
:100030001F192006FF3E08001F171F0D06398C00F0
:100040002B208D008C0A2B201F1F7E281F138C0055
:100050000023095207E2800308A000C088207013487
:10006000753403341534003 43C340C34D9348F00E7
:100070002208840020095E2084138F0803199A281F
:10008000F03091000E0880389000F0309103031991
:10009000910003198F0303199A284E286120030148
:1000A0000C1820088E1F20088E0803190301900FDA
:1000B0005B28800642285C280000452884178005BC
:1000C0009A280D080C0403198C0A80300C1A8D062E
:1000D0000C198D068C188D060D0D8C0D8D0D9A2822
:1000E0008F018E00FF308E07031C8F07031C9A2898
:1000F00003308D00DF307E2072288D01E83E8C00B9
:100100008D09FC30031C87288C07031884288C0772
:1001100064008D0F84280C188D288C1C91280000F9
:10012000912808000 3108D0C8C0CFF3E03189228B8
:100130000C089A288313031383 12640008008316A3
```

PROGRAM 6.12

rx-remote.hex file listing

PROGRAM 6.12

rx-remote.hex file listing
(continued)

```
:10014000850101308600831206128316061283127F
:100150000613831606138312861283168612831  2E1
:10016000861083168610831206118316061183122D9
:1001700086118316861105308312A2000830A00074
:1001800073308E000A30372032308E000A3037202C
:100190000630A2000130A00004309F0001205A3C2C
:1001A000031DCE280120A40064002408413C031D47
:1001B000DA281B2164002408423C031DE0287D212D
:1001C00064002408433C031DE6284C2164002408F5
:1001D000443C031DEC28AE2164002408453C031D6B
:1001E000FD280530A2000830A00073308E000A30D0
:1001F000372032308E000A30372064002408463C15
:10020000031D1A2906128316061283120613831 67B
:10021000061383128612831686128312861083166A3
:10022000861083120611831606118312861183161  7
:1002300086118312C82806168316061283120617 23
:100240008316061383120130  8F002C307120640056
:100250000  5182B292729061283160612831206136  6
:1002600083160613831286148316861083120615CE
:1002700083160611831201308F002C307120640028
:10028000851843293F2986108316861083120  6118C
:10029000831606118312080006168316061283122AF
:1002A00086168316861283120130  8F002C307120 3F
:1002B0006400051  85C29582906128316061283125 9
:1002C00086128316861283128614831686108312 72
:1002D0000615831606118312013  08F002C30712011
:1002E00064008518742970298610831686108312 7D
:1002F0000611831606118312080006168316061  2CD
:100300008312861683168612831201308F002C30DA
:1003100071206400051  88D298929061283160612 9A
:10032000831286128316861283128614831686101 1
:1003300083128615831686118312013  08F002C30AC
:10034000712064008518A529A129861083168610BE
:1003500083128611831686118312080006168316EF
:10036000061283120617831606138312013  08F00BC
:100370002C30712064000518BE29BA290612831 94
:100380000612831206138316061383128614831  2D
:10039000861083128615831686118312013  08F001  2
```

```
:1003A0002C30712064008518D629D2298610831636
:1003B000861083128611831686118312080063004B
:0203C000DF2933
:02400E00F53F7C
:00000001FF
```

PROGRAM 6.12

rx-remote.hex file listing
(continued)

PROGRAM 6.13

tx-remote.bas program
listing

```
'_____

' Name      : tx-remote.bas
' Compiler  : PicBasic Pro - MicroEngineering Labs
' Notes     : Robot control using the Linx 433LC series
'           : transmitter and receiver.
'           : Using the PIC 16C71 on-chip analog to digital
'           : converters to read the position of
'           : the two control stick potentiometers.
'_____


' PIC 16C71 A/D converter registers
'
' PORTA = 05 hex = 5 dec
' five I/O lines RA0 RA1 RA2 RA3 RA4
'
' TRISA = 85 hex = 133 dec
' data direction register
' —1 1111 inputs
' —0 0000 outputs
'
' ADCON1 = 88 hex = 136 dec
' configure as A to D converter or digital I/O
'    bits   RA0,RA1   RA2      RA3      Vref
' — –00  analog    analog   analog   VDD
' — –01  analog    analog   ref input RA3
' — –10  analog    digital  digital  VDD
' — –11  digital   digital  digital  VDD
'
' ADCON0 = 08 hex = 8 dec
' A/D control and status register - 8 bits
' bit7 - ADCS1
' bit6 - ADCS0
```

PROGRAM 6.13

tx-remote.bas program listing (continued)

```
' bit5 - reserved
' bit4 - CHS1
' bit3 - CHS0
' bit2 - GO/DONE
' bit1 - ADIF
' bit0 - ADON
' ADCS1 and ADCS2 - bit7 and bit6
' A/D conversion clock select:
' ADCS1,0 =    00:      fosc/2
'              01:      fosc/8
'              10:      fosc/32
'              11:      f rc (derived from internal
' rc oscillator)
' bit5 - reserved
' Analog channel select - bit4 and bit3
' CHS1, CHS0 = 00: channel 0 (AIN0)
'              01: channel 1 (AIN1)
'              10: channel 2 (AIN2)
'              11: channel 3 (AIN3)
' GO/DONE - bit2: must be set to begin a
'                 conversion. It is automatically
'                 reset in hardware when conversion
'                 is done.
' ADIF - bit1:    A/D conversion complete interrupt flag bit. Set
'                 when conversion is completed. Reset in software.
' ADON - bit0:    If ADON = 0 A/D converter module is shut off and
'                 consumes no operating current. ADON = 1 A/D
'                 converter module is on.
'
' ADRES = 09 hex = 9 dec
' A/D conversion result register
'
' INTCON = 0B hex = 11 dec
' interupt control register


'_____

' set PortA inputs.
trisa = %00011111
```

PROGRAM 6.13

tx-remote.bas program
listing (continued)

```
' PortB set as outputs. Pin 2 input
trisb = %00000100


'_____


' initialize variables

include "modedefs.bas"

tx_baud        CON N2400
pot_y          VAR PORTA.0
pot_x          VAR PORTA.1
txmit          VAR PORTB.0
txmit_led      VAR PORTB.1
push_button    VAR PORTB.2
val_y          VAR BYTE
val_x          VAR BYTE
control        VAR BYTE


'_____


' Set up the analog to digital converters

DEFINE ADC_BITS 8          ' Set number of bits in result
DEFINE ADC_CLOCK 3         ' Set clock source (rc = 3)
DEFINE ADC_SAMPLEUS 10     ' Set sampling time in microseconds
ADCON1 = 2                 ' Set porta pins 0 and 1 to analog

start:

low txmit_led

ADCIN 0,val_y      ' read A/D converter - porta.pin 0
ADCIN 1,val_x      ' read A/D converter - porta.pin 1

If val_y < 20 then
    high txmit_led
    serout txmit,tx_baud,["ZA"]
```

PROGRAM 6.13

tx-remote.bas program
listing (continued)

```
endif

If val_y > 200 then
    high txmit_led
    serout txmit,tx_baud,["ZB"]
endif

If val_X < 20 then
    high txmit_led
    serout txmit,tx_baud,["ZC"]
endif

If val_X > 200 then
    high txmit_led
    serout txmit,tx_baud,["ZD"]
endif

If push_button = 1 then
    high txmit_led
    serout txmit,tx_baud,["ZE"]
endif

If ((val_y > 25) and (val_y < 190)) or ((val_x > 25) and (val_x < 190))
then
    serout txmit,tx_baud,["ZF"]
endif

goto start

end
```

PROGRAM 6.14

tx-remote.hex file listing

```
:100000008C2892002A0884000930930003100D20E8
:10001000920C930B072803140D288413A71D1C288A
:1000200000082804271D2806800084170008280 4DB
:10003000031C2806800027280008280403 1C280623
:10004000271928068000841728098005272827 0DEE
:100050006398C0030208D008C0A302000004E289C
:1000600000308A000C08820701347534033415 34DB
```

```
:1000700000343C340C34D9348C008C0D8C0D0C0DB8
:100080003839C138880000308D000A304E200815FC
:10009000081948288D01090887288D01E83E8C0041
:1000A0008D09FC30031C57288C07031854288C0733
:1000B00064008D0F54280C185D288C1C61280000EA
:1000C000612808008D018F018E0001306C288D01A0
:1000D0008F018E0004306C2894000F080D02031D60
:1000E00073280E080C020430031801300319023083
:1000F0001405031DFF3087280038031DFF30040559
:10010000031DFF3087280404031DFF308728831355
:10011000031383126400080083161F308500043027
:100120008600023088008312861083168610003005
:1001300083123C20AE0001303C20AD00640014303E
:100140002E020318B12886148316861006308312F7
:10015000AA000130A8000430A7005A300120413025
:100160001206400C9302E02031CC42886148316A3
:10017000861006308312AA000130A8000430A700C0
:100180005A30012042300120640014302D0203183F
:10019000D72886148316861006308312AA000130F1
:1001A000A8000430A7005A30012043300120640029
:1001B000C9302D02031CEA288614831686100630E7
:1001C0008312AA000130A8000430A7005A30012091
:1001D000443001206400061DFB2886148316861017
:1001E0006308312AA000130A8000430A7005A305C
:1001F000120453001202E088C00193062209E001D
:100200002E088C00BE306720A0001E088400200845
:100210007C20A000A1002D088C0019306220A200D3
:100220002D088C00BE306720A4002208840024081A
:100230007C20A400A5020082104840024082504B3
:100240008320A400A5006400240825040319322992
:100250000630AA000130A8000430A7005A3001205F
:0602600046300120942845
:02400E00FD3F74
:00000001FF
```

7

Turtletron: Build Your Own Robotic Turtle

Turtles and Tortoises

There are more than 270 living species of turtles and tortoises. These creatures are found in terrestrial, fresh water, and marine habitats, and in both temperate and tropical regions. The term "turtle" usually refers to a freshwater or marine species, while the term "tortoise" is normally used for terrestrial species. "Terrapin" is the informal name for a freshwater turtle.

Turtles and tortoises belong to the order Testudines, which is divided into two suborders. The primitive sideneck turtles (suborder Pleurodira) cannot fully retract their long necks. When they are at rest, they must lay their heads sideways along the inside of their shells. All of the 70 or so species of sideneck turtles live in freshwater. The more advanced straightneck turtles (suborder Cryptodira) are a much larger group that lives on land and in water. They are able to withdraw their heads completely into their shells.

Turtles and tortoises vary greatly in size, from the tiny Speckled Padloper, 2-1/2 inches long, to the massive Leatherback Sea Turtle, which can reach up to 6 feet in length.

FIGURE 7.1

A turtle and its robotic counterpart.

The turtle and its behavior is the inspiration for the robot in this chapter. At first I wanted the turtle to be a walking robot, much like the biological version, but decided that an inexpensive, wheeled robot would be a great platform on which to base experiments. **Figure 7.1** shows a real turtle and the robotic version that will be built during this chapter.

Overview of the Turtletron Project

The robot turtle that will be built and programmed in this chapter has a circular base and achieves locomotion using two wheels, each one powered by direct current (DC) motors and gearboxes. The robot will operate in autonomous mode or under remote control by a human operator. Turtletron will use an ultrasonic range finder and a linear shaft encoder to map its surrounding area during autonomous mode, and will also use the sonar to inhibit movement if an operator is directing the robot into an obstacle during remote control. The robot will also be equipped with a linear shaft encoder that will give it the ability to keep track of the distance that

FIGURE 7.2

Turtletron with remote control.

it has traveled and to create maps of its surroundings. To save time and money on construction, this robot will use the same main controller circuit board and transmitter device that we built during the last crocodile robot project. The only difference with the main controller board will be with the software of the PIC 16F84. The robot will also adopt the wireless data link that was utilized in the last chapter. The robot with the remote control is shown in **Figure 7.2**.

The History of Robotic Turtles

William Grey Walter built the first robotic turtles in the late 1940s. His work in robotics was an extension of his research in neurophysiology. Walter's studies of the brain and its neural networks led him to wonder about what type of behavior could be created using just a few neurons. To experiment with this concept, in 1948, Walter built a three-wheeled turtle-like mobile robot that measured 12 inches in height and 18 inches in length. Amazingly this robot used just two electronic neurons, but exhibited interesting and complex behaviors. The first two robots were named Elmer and Elsie (ELectroMEchanical Robot, Light Sensitive). He later named the style of robots Machina Speculatrix after observing the complex behavior they exhibited.

The robot's nervous system consisted of two sensors connected to two neurons. One sensor was a light-sensitive resistor mounted onto the shaft of the front wheel steering-drive assembly. This arrangement ensured that the photosensitive resistor was always facing in the direction that the robot was moving. The second sensor was a bump switch attached to the robot's outer cover. The three wheels of the robot were arranged in a triangular configuration. The front wheel had a motorized steering assembly that could rotate a full 360 degrees in one direction. The front wheel also contained a drive wheel for propulsion. **Figure 7.3** shows a robot turtle built by Walter during the 1940s. This robot is now on display at the Smithsonian.

FIGURE 7.3

Robot tortoise built by robotics pioneer William Grey Walter in 1948.

The robot exhibited four modes of operation described below.

1. **Search.** The room is at low light level or darkness. The robot responds by searching for a light source. The steering motor is on full speed and the drive motor is at half speed.

2. **Move.** The robot found light. The robot responds by turning the steering motor off and the drive motor on at half speed.

3. **Dazzle.** The robot encounters bright light. The robot responds by setting the steering motor to half speed, while the drive motor is reversed.

4. **Touch.** The robot hits an obstacle. The robot responds by setting the steering motor to full speed, with the drive motor reversed.

In the 1950s, W. Grey Walter wrote two *Scientific American* articles ("An Imitation of Life," May 1950; "A Machine That Learns," August 1951) and a book titled *The Living Brain* (Norton, New York, 1963). Walter reported, "The strange richness provided by this particular sort of permutation introduces right away one of the aspects of animal behavior—and human psychology—that Machina Speculatrix is designed to illustrate: the uncertainty, randomness, free will or independence so strikingly absent in most well designed machines."

Although the robot we will be building is turtle-like, it is not intended to recreate any of the experiments of W. Grey Walter, although you could easily implement the sensors and program the microcontroller to do so.

Mechanical Construction of Turtletron

The parts needed for the mechanical construction of the turtle robot are listed in **Table 7.1**.

TABLE 7.1	**Parts**	**Quantity**
List of Parts Needed for Turtletron's Mechanical Construction	18 3/4-inch diameter Frisbee	2
	3-inch diameter model airplane wheels	2
	1-1/2 inch casters	2
	#4-40 × 3/4-inch machine screws	4
	#4-40 × 1-inch machine screws	4
	#4-40 nuts	8
	6/32 × 1/2-inch machine screws	32
	6/32 × 1-inch machine screws	2
	6/32 locking nuts	34
	Power switch DPDT	1
	Tamiya high power gear box H.E.	2
	Connector wire	9 feet
	Heat-shrink tubing	2 inches
	4-post female header connector	3

The construction of the robot turtle will start with the assembly of two Tamiya high power gearboxes. They are available from HVW Tech and can be purchased at their Web site, located at www.hvwtech.com. The gearboxes are sold as kits and need to be assembled before they can be used. **Figure 7.4** shows the Tamiya high power gearbox kit.

FIGURE 7.4

Tamiya high power
gearbox kit.

Assembling the Gearboxes and Attaching the Wheels

Take all of the parts out of the box and unfold the instruction sheet. The gearbox has two possible configuration options of a 64.8:1 or 41.7:1. The gearbox will be assembled for use with the 64.8:1 ratio using one green and two red gears. Follow the instructions included with the kits to assemble both gearboxes.

Locate two, 3-inch diameter model airplane wheels and two gearbox horns labeled as A3 that are included with the gearbox kits. Place a wheel on the table and line up the center hole in one of the gearbox horns with the center of the wheel. Use a pencil to mark

FIGURE 7.5

Gearbox horn A3 with mounting holes indicated.

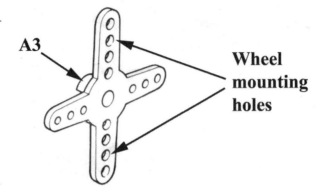

the position of the two holes where they line up on the wheel, as shown in **Figure** 7.5. Follow this procedure for the second wheel,

FIGURE 7.6

Wheel attached to gearbox.

and then drill the holes with a 5/32-inch drill bit. Attach an A3 gearbox horn to each of the gearboxes with the washers and securing nuts that came with the kits. Use two #4-40 × 3/4-inch machine screws and nuts to attach each wheel to each gearbox horn as, shown in **Figure 7.6**.

Constructing the base.   The robot's body will be constructed using two common Frisbees that can be obtained at most department stores. Starting with the base, use the dimensions shown in **Figure 7.7** to cut two recesses in the plastic disk, using a hack saw. The gearboxes will be mounted so that the wheels are positioned in the recessed areas. Use a file to smooth any rough edges where the plastic was cut.

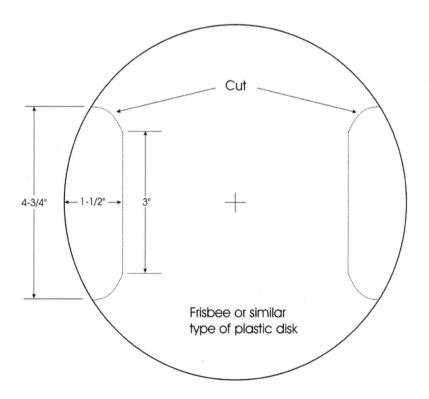

FIGURE 7.7

Cutting dimension for wheel recesses in the robot base.

Cut

4-3/4"   1-1/2"   3"

Frisbee or similar type of plastic disk

Drill the motor mounting holes and power switch hole, as indicated in **Figure** 7.8. Center the casters at the front and back of the underside of the Frisbee, and mark the mounting holes with a pencil, as shown in **Figure** 7.8. No dimensions for drilling were shown in the figure because the exact position of the caster mounting holes may vary, depending on the casters. When the holes have been marked, drill with the bit sizes indicated.

FIGURE 7.8

Drilling guide for robot base.

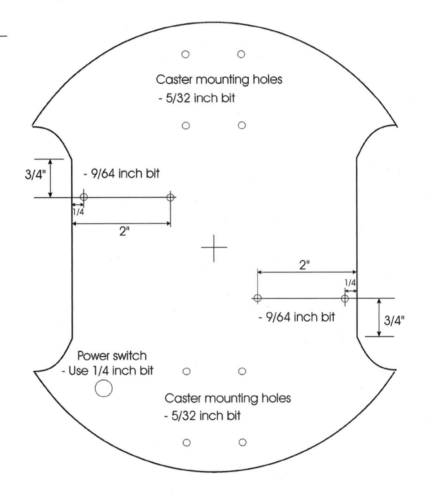

Mount the wheeled gearboxes onto the robot base using the machine screws and nuts that came with the gearbox kits. Mount each caster onto the base using four 6/32-inch × 1/2-inch machine screws and locking nuts. Mount the power switch in the 1/4-inch hole toward the back of the base. Use **Figure 7.9** to position the gearboxes, casters, and switch when mounting them to the base.

FIGURE 7.9

Gearboxes, wheels, casters, and switch mounted to the robot base.

Cut four pieces of 1/2-inch aluminum stock to a size of 2-1/2 inches in length. These pieces will be used to support the top cover and antenna. Use **Figure 7.10** as a cutting and drilling guide. Mount the aluminum pieces on the Frisbee base. Position each piece 1/2 of an inch beside the wheel recesses and mark the mounting holes. Drill each mounting hole with a 5/32-inch drill bit, and attach each piece with a 6/32-inch × 1/2-inch machine screw and locking nut. **Figure 7.11** shows two of the supports attached around one of the wheel recesses.

FIGURE 7.10

Cutting and drilling guide for cover supports.

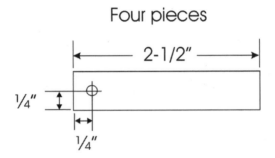

FIGURE 7.11

Two cover supports mounted to robot base around wheel recesses.

Differential drive system. Turtletron employs what is called the differential drive system. It is one of the least complicated locomotion systems from a construction and programming standpoint. The differential drive scheme consists of two wheels on a common axis, with each wheel driven independently. This arrangement allows the robot to drive straight, to turn in place, and to move in an arc.

In order to ensure balance, some additional support beside the two drive wheels must be provided to prevent the robot from tipping over. This is usually done by arranging one or two caster wheels in a diamond or triangle pattern. Turtletron uses the diamond pattern, as illustrated in **Figure** 7.9. One of the problems with using this configuration is that when the caster wheels are attached rigidly to the robot body, undulations in terrain can leave the robot supported only by the casters. The drive wheels may lose contact with the surface and become unable to move the robot. To improve on this design, a suspension system could be added that would allow the casters to move up and down relative to the drive wheels.

Electronics

To simplify the design and construction of Turtletron, the main controller board and remote control that were built for the crocodile robot in the last chapter will be used. The circuits are identical, except that the software of the PIC 16F84 will be changed. This robot will also include an ultrasonic range finder for room mapping and obstacle avoidance, along with a linear shaft encoder to keep track of distance. The main controller schematic is shown in **Figure** 7.12. If you did not build the crocodile robot, or would like to build a separate circuit board for Turtletron, follow the instructions in Chapter 6. The parts needed to complete the electronics are listed in **Table** 7.2.

FIGURE 7.12

Schematic of
Turtletrons main
controller board.

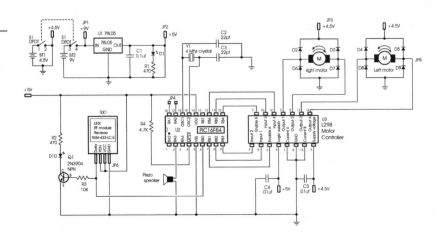

TABLE 7.2

List of Parts Needed for
Turtletron's Electronics

Part	Quantity	Description
Semiconductors		
U1	1	78L05 5V regulator
U2	1	PIC 16F84 flash microcontroller mounted in socket
U3	1	L298 dual full-bridge driver
RX1	1	Lynx RXM-433-LC-S RF receiver module
D1	1	Red light-emitting diode
D2–D9	8	Diodes 1N4001
D10	1	Green light-emitting diode
Q1	1	2N3904 NPN transistor
Resistors		
R1, R2	2	470 Ω 1/4-watt resistor
R3	1	10 KΩ 1/4-watt resistor
R4	1	4.7 KΩ 1/4-watt resistor

(continued on next page)

Part	Quantity	Description	
Capacitors			**TABLE 7.2**
C1	1	0.1 µf	List of Parts Needed for Turtletron's Electronics (continued)
C2, C3	2	22 pf	
C4, C5	2	.01 µf	
Miscellaneous			
JP1–JP4	4	2-post male header connector—2.5-mm spacing	
JP5–Motors	1	4-post male header connector—2.5-mm spacing	
JP6–RF module	1	4-post female header connector—2.5-mm spacing	
Y1	1	4-MHz crystal	
W1-W4	4	Jumper wire	
Piezo buzzer	1	Standard piezoelectric element	
I.C. socket	1	18-pin I.C. socket—soldered to PC board U2	
Whip antenna	1	6-3/4 inch whip antenna	
9-volt battery strap	2	Battery connector	
4 AA-battery holder	1	4 AA-battery holder with 6-volt output	
Printed circuit board	1	See details in Chapter 6.	

Mount the main board on four 1/2-inch threaded standoffs. Turn the robot over so that it is right side up, with the wheels facing downward. Place the main controller circuit board at the center of the robot base, mark the positions of the standoffs, and then drill the holes with a 5/32-inch bit. Mount the circuit board to the robot base using machine screws that match the threaded standoffs.

Ultrasonic Range Finding

An ultrasonic range finder will be added to Turtletron so that the robot will be able to avoid obstacles while roaming in autonomous mode or to inhibit movement when under remote control. The robot will be able to determine the distance to an object from itself, and then make decisions based on that information. The robot will also have the ability to create a rudimentary map of the surrounding area before movement through the environment begins.

Devantech SRF04 ultrasonic range finder.   A low-cost solution is the Devantech SRF04 ultrasonic range finder, pictured in **Figure 7.13**. This device offers precise ranging information from 3 cm to 3 m, is easy to interface, and its minimal power requirements make it an ideal ranger for mobile robotics applications. It is available from Acroname Robotics Inc., and can be purchased from their Web site at www.acroname.com.

FIGURE 7.13

Devantech SRF04 ultrasonic range finder.

The SRF04 range finder is a small printed circuit board (PCB) that measures 1-3/4 × 3/4-inches, with two ultrasonic transducers mounted on the front. The ranger requires a 5V power supply capable of handling roughly 50 mA of continuous output. One transducer is used to send an ultrasonic signal, and the other transducer receives the signal reflection from nearby objects. The SRF04 will output a 100-microsecond to 18-millisecond detection pulse that is proportional to range when a reflected signal is detected. **Table 7.3** is a list of the parts that will be needed to add the sonar ranger. The SRF04 range finder specifications are listed in **Table 7.4**.

Part	Quantity	Description
SRF04 ultrasonic ranger module	1	Sonar distance measuring device
5-post male header connector	1	2.5-mm spacing
4-strand ribbon cable	1	8-1/2 inches
2-connector female header	2	2.5-mm spacing
1/16-inch thick aluminum	1	2-inches × 4-inches
Hot glue	—	Hot glue and gun

TABLE 7.3

Parts Required for the Addition of the SRF04 Ultrasonic Range Finder

Specification	Value
Voltage	5v
Current	30 mA Typical 50 mA
Frequency	40 kHz
Maximum range	3 meters
Minimum range	3 centimeters
Sensitivity	Can detect a 3-cm diameter broom handle at 2 meters
Input trigger	10 µS minimum TTL level pulse
Echo pulse	Positive TTL level signal, width proportional to range
Size	1-3/4 × 3/4-inches

TABLE 7.4

Table of Specifications for the SRF04

Theory of operation. The SRF04 works by sending a pulse of sound outside the range of human hearing. This pulse travels at the speed of sound (1.1 ft/ms) away from the ranger in a cone shape. If any objects are in the path of the pulse, the sound is reflected off the object and back to the ranger. The ranger is paused for a brief interval after the sound is transmitted and then awaits the reflected sound in the form of an echo. The controller driving the ranger requests the device to create a 40-kHz sound pulse, and then waits for the return echo. If the echo is received, the ranger reports this echo to the controller, and the controller can then compute the distance to the object, based on the elapsed time.

Connections. The ranger requires four connections to operate. The first two are the power and ground lines. The ranger requires a 5-volt power supply capable of handling roughly 50 mA of continuous output. The other two lines are the signal connections. The first signal connection is the pulse trigger input line, and the second is the echo output line. These two pins will be connected to two input/output (I/O) lines of the microcontroller. **Figure 7.14** shows the connection pins on the back of the device. Note that the ground pin is on the far right and is marked with the letter "G" beside it.

FIGURE 7.14

SRF04 pin connections.

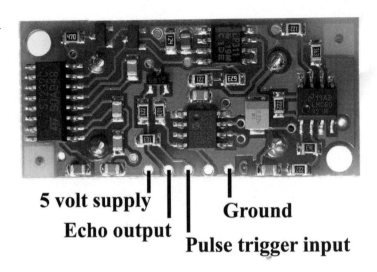

5 volt supply

Echo output

Ground

Pulse trigger input

Basic timing. There are a couple of requirements to consider about the input trigger and the output pulse generated by the ranger. The input line should be held low (logic 0), and then brought high for a minimum of 10 µsec to initiate the sonic pulse. The pulse is generated on the falling edge of the input trigger. The ranger's receive circuitry is held in a short blanking interval of 100 µsec to avoid noise from the initial ping, and then it is enabled to listen to the echo. The echo line is logic low until the receive circuitry is enabled. Once the receive circuitry is enabled, the falling edge of the echo line signals either an echo detection or the time-out of 36 ms if no object is detected. **Figure 7.15** illustrates the timing sequence of the initial trigger input, the 40 kHz sonic burst that is generated, and the echo output pulse.

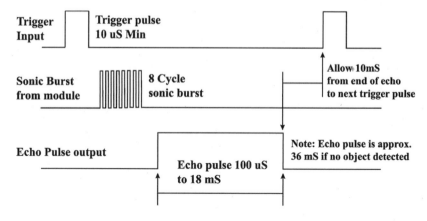

FIGURE 7.15

SRF04 timing diagram.

The microcontroller will begin timing on the falling edge of the trigger input pulse, and end timing on the falling edge of the echo line. This duration determines the distance between the sonar module and the object from which the echo is bounced back. If no object is detected, a time-out will occur, which is indicated by the echo output line going high for approximately 36 ms.

FIGURE 7.16

Header pins soldered to
the SRF04 ultrasonic
ranger.

Header pins soldered to board

Connecting the ultrasonic ranger to the robot. First, solder four male header pins to the ranger, as shown in **Figure 7.16**. This is probably the best way to connect the ranger to the controller because the robot could possibly move the wires around during locomotion. Wires soldered directly to PCBs have a tendency to fray at the solder joints and become disconnected. The use of header pins eliminates this problem.

Fabricate a jumper wire made up of 4-strand ribbon wire cut to a length of 8-1/2 inches. The end of the wire attached to the ranger uses a 5-connector female header. Solder the wires to the female header connector. Skip the pin that is not used and clip it off with wire cutters. On the other end of the wire, use a pair of 2-connector female headers and solder the 5-volt and ground to one connector, and the trigger input and echo output to the other. **Figure 7.17** illustrates what the connector wire should resemble when it is finished.

FIGURE 7.17

Completed SRF04 connector wire.

To secure the ultrasonic ranger to the robot, a housing to mount the unit will be fabricated. Use **Figure 7.18** as a guide to cut, drill, and bend the housing, using 1/16-inch thick aluminum. Drill the mounting hole with a 5/32-inch drill bit. The aluminum can be bent on the edge of a table by hand or in a table vise. **Figure 7.19** shows the finished housing so that you can get an idea of how the aluminum should be bent. Next, place the ranger unit inside the housing at the front and secure it in place by tightening the aluminum around the circuit board by hand. Apply a small amount of hot glue on the inside at the corners where the circuit board and aluminum housing meet. This will ensure that the circuit board does not move out of position.

FIGURE 7.18

Cutting, drilling, and bending guide for the SRF04 housing.

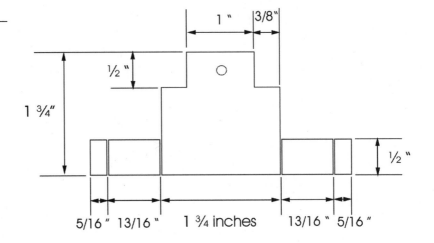

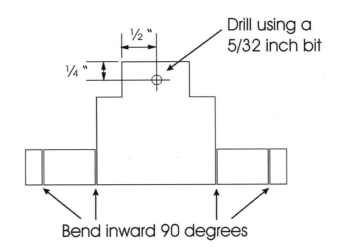

FIGURE 7.19

Finished SRF04
housing.

Figure 7.20 shows the SRF04 ranger mounted in the housing with
the jumper wire plugged into the header connector. Use 1/16-inch
thick aluminum stock to construct the neck mount that will connect the ranger housing to the robot body. Follow the cutting,
bending, and drilling guide in **Figure 7.21**. When the neck mount
is completed, attach it to the front of the robot with a 6/32-inch ×
1/2-inch machine screw and locking nut, as shown in **Figure 7.22**.
Note that the 1-1/2 inch section of the neckpiece is attached to the
robot base. Attach the ultrasonic ranger housing to the neck using
a 6/32-inch × 1/2-inch machine screw and locking nut.

FIGURE 7.20

SRF04 ranger mounted
in housing with
connector wire
attached.

FIGURE 7.21

Cutting, bending, and
drilling guide for neck
mount.

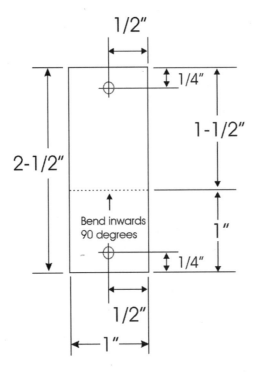

1/2"

2-1/2"

1/4"

1-1/2"

Bend inwards
90 degrees

1"

1/4"

1/2"

1"

holes drilled with a 5/32 inch bit

FIGURE 7.22

Neck mount and sonar ranger attached to Turtletron's body.

Attaching the antenna to the RF module. Locate the 6-3/4 inch whip antenna and strip 1/2-inch of the insulator and shielding material from the connector wire. Drill a hole in the second Frisbee, toward the edge, using a 1/4-inch bit. Mount the whip antenna to the Frisbee by feeding the connector lead through the hole and then fastening the mounting nut. Solder the wire to the antenna mount area on the back of the Lynx RXM-433-LC-S receiver module. Bend the pins on the receiver module 90 degrees downward, if this was not done earlier in Chapter 6. The finished top cover with the antenna and receiver module attached is shown in **Figure** 7.23.

FIGURE 7.23

Antenna and receiver module attached to top cover.

Now that all of the components are in place, it is time to wire everything together. Use the diagram in **Figure 7.24** to connect all of the components to the main controller board. Drill a 5/32-inch hole in the base in front of each of the motors to feed the motor wires through to the controller board. Plug the RF receiver module into the 4-connector female header on the controller board. Attach the top cover with the antenna toward the back of the robot. The top cover should fit snugly on the four aluminum cover support pieces. **Figure 7.25** shows the robot with all of the components and batteries connected to the main controller board. Attach a fresh 9-volt battery and a 6-volt battery pack containing four AA batteries to the proper battery clips, as indicated in **Figure 7.24**.

In the next section, we will program the PIC 16F84 to control the motors, interpret the information from the radio receiver module, and obtain distance measurements from the sonar ranger for obstacle avoidance and room mapping. The final experiment will be to add an optical shaft encoder so that the robot will be able to keep track of the distance that it has traveled. This will also be necessary when the robot is creating maps of its surrounding environment.

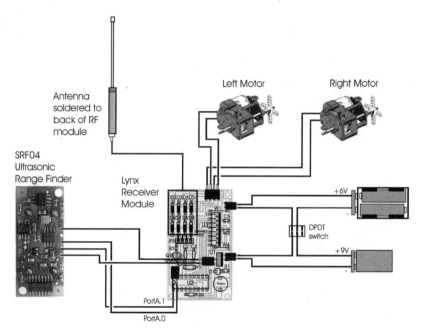

FIGURE 7.24

Turtletron wiring diagram.

FIGURE 7.25

Turtletron with all
components attached.

The Remote Control Transmitter

The first objective will be to control Turtletron's differential drive,
using the remote control transmitter that was built in Chapter 6.
The hand held remote control device uses an analog X and Y axis
control stick as the input to two analog-to-digital converters residing on a PIC 16C71. To make the project easier, we will not change
any of the programming for the remote control transmitter. If you
wish to create another remote control, follow the instructions in
chapter 6. To make Turtletron respond only to the second remote
control, simply change the qualifier in the serial transmit and
receive code of the robot and transmitter. The schematic for the
transmitter remote control is shown in **Figure 7.26**.

The circuit functions by using a PIC 16C71 to monitor the position
of the analog control stick and then send serial commands to the
transmitter module. When the control stick moves along the X and
Y axis, the resistance values of two 100KΩ potentiometers change
proportionally. The control stick and the two attached poten-

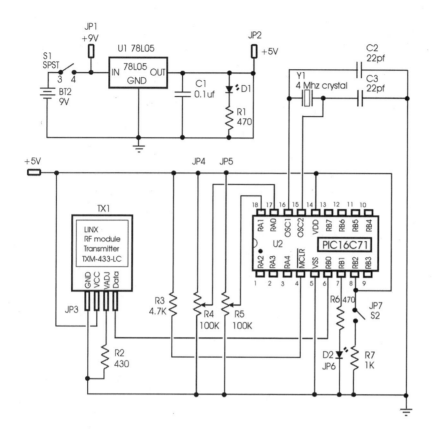

Figure 7.26

Remote control
schematic diagram.

tiometers are shown in Chapter 6 (**Figure 6.47**). Each potentiometer is configured as a voltage divider, so that a unique voltage represents each position along the X and Y axis. The voltages
from the potentiometers are converted to 8-bit values by the internal analog-to-digital converters on the PIC 16C71, and then interpreted by the microcontroller. Depending on the values, certain
movement commands are sent in a serial format from the transmitter to the robot. The remote control also has a programmable
push-button switch and a light-emitting diode (LED) that can be
turned on when certain events occur, such as during the transmission of a movement command. The transmitter module is the
TXLC-434 transmitter, available from Reynolds Electronics at
www.rentron.com.

Programming Turtletron

The first program to be written will receive commands from the hand held remote control via the RF receiver module. This information will be used to control the drive motors, as required. It will not be necessary to reprogram the transmitter because the same transmission codes that were implemented for the final remote control program in Chapter 6 will be used. The robot control program will use the serin command to collect the data from the receiver module, and then make movement decisions based on that information. The differential drive allows the robot to move forward, reverse, turn left, or turn right on the spot, and to move in an arc. The control program is called turtle-receive.bas and is listed in **Program 7.1**. Compile turtle-receive.bas ,and then program the PIC 16F84 with the turtle-receive.hex file listed in **Program 7.2**. Place the PIC 16F84 into the 18-pin socket on Turtletron's main board.

If you reprogrammed the PIC 16C71 in the transmitter circuit since Chapter 6, then compile turtle-trans.bas listed in **Program 7.3**. Program the 16C71 with the turtle-trans.hex file listed in **Program 7.4**, and then insert the PIC back into the 18-pin socket on the transmitter circuit board. Move the control stick on the remote control to the middle position, and then turn the power on. Turn the robot on and place it on the ground. When the control stick is moved to the forward position, the robot will move forward. With the stick moved backwards, the robot will respond by moving in reverse. With the control stick moved to the left, the robot will rotate left on the spot. The ability to rotate on the spot is one of the great things about using a differential drive system. Rotating on the spot is accomplished by rotating one wheel forward, while the other wheel rotates in reverse. With the stick moved to the right, the robot will rotate to the right on the spot. Try moving the control stick to the forward-right position. The code will alternate transmitting forward and turn-right commands to the robot. The robot will respond by moving in a forward–right arc.

PROGRAM 7.1

turtle-receive.bas listing

```
'_____
' Name     : turtle-receive.bas
' Compiler : PicBasic Pro - MicroEngineering Labs
' Notes    : Robot remote control using the Linx 433LC
'          : series transmitter and receiver.
'_____


' PortA set as outputs.
trisa = %00000000

' PortB set as outputs. pin 0 input.
trisb = %00000001


'_____

' initialize variables

include "modedefs.bas"

rx_baud        CON N2400
rxmit          VAR PORTB.0
enable_right   VAR PORTB.1
forward_right  VAR PORTB.2
reverse_right  VAR PORTB.3
enable_left    VAR PORTB.4
reverse_left   VAR PORTB.5
forward_left   VAR PORTB.6
piezo          VAR PORTA.3
control        VAR BYTE
temp           VAR BYTE

low enable_left
low forward_left
low reverse_left

low enable_right
```

PROGRAM 7.1

turtle-receive.bas listing
(continued)

```
low forward_right
low reverse_right

SOUND PIEZO,[115,10,50,10]

start:

serin rxmit,rx_baud,["Z"],control

if control = "A" then
   gosub forward
endif

if control = "B" then
   gosub backwards
endif

if control = "C" then
   gosub turn_left
endif

if control = "D" then
   gosub turn_right
endif

if control = "E" then
   sound piezo,[115,10,50,10]
endif

if control = "F" then
   low enable_left
   low forward_left
   low reverse_left
   low enable_right
   low forward_right
   low reverse_right
endif
```

```
goto start
'_____

' movement subroutines

forward:

    high enable_left
    high forward_left

    high enable_right
    high forward_right


return


'_____


turn_left:

    high enable_left
    high forward_left

    high enable_right
    high reverse_right


return


'_____


backwards:

    high enable_left
    high reverse_left

    high enable_right
    high reverse_right


return
```

PROGRAM 7.1

turtle-receive.bas listing
(continued)

PROGRAM 7.1

turtle-receive.bas listing
(continued)

```
'_____

turn_right:

    high enable_left
    high reverse_left

    high enable_right
    high forward_right

return

end
```

PROGRAM 7.2

turtle-receive.hex file
listing

```
:100000009228640011200318012B1C2008308F005A
:100010001D2011208E0C0C288F0B08281D200E0887
:1000200008002208840020088417800484130005 37
:100030001F192006FF3E08001F171F0D06398C00F0
:100040002B208D008C0A2B201F1F71281F138C0062
:1000500002308820712800308A000C0882070134A1
:100060007534033415340034 3C340C34D9348F00E7
:100070002208840020095E2084138F0803198D282C
:10008000F03091000E0880389000F0309103031991
:100090000910003198F0303198D284E286120030155
:1000A0000C1820088E1F20088E0803190301900FDA
:1000B0005B28800642285C280000452884178005BC
:1000C0008D280D080C0403198C0A80300C1A8D063B
:1000D0000C198D068C188D060D0D8C0D8D0D8D282F
:1000E0008D01E83E8C008D09FC30031C7A288C07BA
:1000F000031877288C0764008D0F77280C18802848
:100100008C1C842800008428080003108D0C8C0CA3
:10011000FF3E031885280C088D28831303138312D0
:100120006400080083160330850001308600831 2C6
:100130006128316061283120613831606138312 01
:100140008612831686128312861083168610831 2F7
:100150006118316061183128611831686110530 47
:100160008312A2000830A00073308E000A303720BE
:1001700032308E000A3037200630A2000130A000 55
```

```
:1001800004309F0001205A3C031DC2280120A40016
:100190006400240B413C031DCE280F21640024087C
:1001A000423C031DD428312164002408433C031D34
:1001B000DA28202164002408443C031DE028422161
:1001C00064002408453C031DF1280530A2000830D6
:1001D000A00073308E000A30372032308E000A3093
:1001E000372064002408463C031D0E29061283169E
:1001F0000612831206138316061383128612836C1
:10020000861283128610831686108312061183166B7
:100210000611831286118316861183128BC280616D6
:100220008316061283120617831606138312861484A
:100230008316861083120615831606118312080092
:100240000616831606128312061783160613831288E8
:100250008614831686108312861583168611831212E0
:1002600008000616831606128312861683168612257
:100270008312861483168610831286158316861C0
:1002800083120800061683160612831286168316BA3A
:10029000861283128614831686108312061583161F
:0A02A000061183120800063005329C1
:02400E00F53F7C
:00000001FF
```

PROGRAM 7.2

turtle-receive.hex file
listing (continued)

```
'_____
' Name      : turtle-trans.bas
' Compiler  : PicBasic Pro - MicroEngineering Labs
' Notes     : Robot control using the Linx 433LC series
'           : transmitter and receiver.
'           : Using on-chip analog to digital converters
'           : of the PIC 16C71 to read the position of
'           : the two control stick potentiometers.
'_____
```

PROGRAM 7.3

turtle-trans.bas listing

```
' set PortA inputs.

trisa = %00011111

' PortB set as outputs. Pin 2 input
```

305

PROGRAM 7.3

turtle-trans.bas listing
(continued)

```
trisb = %00000100

'_____

' initialize variables

include "modedefs.bas"

tx_baud        CON N2400
pot_y          VAR PORTA.0
pot_x          VAR PORTA.1
txmit          VAR PORTB.0
txmit_led      VAR PORTB.1
push_button    VAR PORTB.2
val_y          VAR BYTE
val_x          VAR BYTE


'_____

' Set up the analog to digital converters

DEFINE ADC_BITS 8           ' Set number of bits in result
DEFINE ADC_CLOCK 3          ' Set clock source (rc = 3)
DEFINE ADC_SAMPLEUS 10      ' Set sampling time in microseconds
ADCON1 = 2                        ' Set porta pins 0 and 1 to analog

start:

low txmit_led

ADCIN 0,val_y     ' read A/D converter - porta.pin 0
ADCIN 1,val_x     ' read A/D converter - porta.pin 1

If val_y < 20 then
   high txmit_led
   serout txmit,tx_baud,["ZA"]
endif

If val_y > 200 then
   high txmit_led
```

```
  serout txmit,tx_baud,["ZB"]
endif


If val_X < 20 then
  high txmit_led
  serout txmit,tx_baud,["ZC"]
endif


If val_X > 200 then
  high txmit_led
  serout txmit,tx_baud,["ZD"]
endif


If push_button = 1 then
  high txmit_led
  serout txmit,tx_baud,["ZE"]
endif


If ((val_y > 25) and (val_y < 190)) or ((val_x > 25) and (val_x < 190))
then
  serout txmit,tx_baud,["ZF"]
endif


goto start


end
```

PROGRAM 7.3

turtle-trans.bas listing
(continued)

```
:100000008C2892002A0884000930930003100D20E8
:10001000920C930B072803140D288413A71D1C288A
:100020000082804271D28068000841700082804DB
:10003000031C28068000272800082804031C280623
:1000400027192806800084172809800527282670DEE
:1000500006398C0030208D008C0A302000004E289C
:1000600000308A000C08820701347534033 41534DB
:1000700000343C340C34D9348C008C0D8C0D0C0DB8
:100080003839C138880000308D000A304E200815FC
:1000900081948288D01090887288D01E83E8C0041
:1000A0008D09FC30031C57288C07031854288C0733
```

PROGRAM 7.4

turtle-trans.hex file
listing

PROGRAM 7.4

turtle-trans.hex file
listing (continued)

```
:1000B00064008D0F54280C185D288C1C61280000EA
:1000C000612808008D018F018E0001306C288D01A0
:1000D0008F018E0004306C2894000F080D02031D60
:1000E00073280E080C0204300318013003190230 83
:1000F0001405031DFF3087280038031DFF30040559
:10010000031DFF3087280404031DFF308728831355
:100110003138312640008008316 1F308500043027
:10012000860002308800831286 1083168610003005
:1001300083123C20AD0001303C20AC006400143040
:100140002D020318B128861483168610063083 12F8
:10015000AA000130A8000430A7005A300120413025
:10016000012 06400C9302D02031CC42886148316A4
:10017000086 1006308312AA000130A8000430A700C0
:100180005A3001204230012064001 4302C02031840
:10019000D7288614831686 1006308312AA000130F1
:1001A000A8000430A7005A30012043300120640029
:1001B000C9302C02031CEA28861483 1686100630E8
:1001C0008312AA000130A8000430A7005A30012091
:1001D00044300120 6400061DFB2886148316861017
:1001E00006308312AA000130A8000430A7005A305C
:1001F000012 0453001202D088C00193062209E001E
:100200002D088C00BE306720A0001E088400200846
:100210007C20A000A1002C088C00 19306220A200D4
:100220002C088C00BE306720A4002208840024081B
:100230007C20A400A500200821048400240825 04B3
:100240008320A400A50064002408250403 19322992
:100250000630AA000130A8000430A7005A3001205F
:0602600046300120942845
:02400E00FD3F74
:00000001FF
```

Testing the SRF04 Ultrasonic Ranger

As described previously, the ranger works by emitting a short burst of sound and then listening for the echo. Under the control of the PICmicro MCU 16F84, the SRF04 will emit an ultrasonic (40 kHz) sound pulse. The pulse travels through the air, hits an object, and then bounces back. Since we know that sound travels through air at approximately 1129 feet per second when the temperature is 21 degrees Celsius, we can accurately determine distance by measuring the amount of time between the transmission of the pulse and the return echo. When the temperature drops, the speed of sound through air slows down. If a temperature sensor was added, an algorithm to determine distance based on the speed of sound through air could take the surrounding temperature into account and adjust for differences.

The PicBasic Pro command called PULSIN returns the round trip echo time in 10 µs units when using a 4-MHz oscillator. Since the pulse width has a 10 µs resolution per increment, that means that if PULSIN returns a value of 1, then 10 µs have elapsed. The factors to convert the raw data to inches and centimeters given in the SRF04 manual are 74 for inches (73.746 µs per 1 inch) and 29 for centimeters (29.033 µs per 1 cm) based on the Basic Stamps PULSIN command returning values in 2 µs increments. In the SRF04 manual, the calculation to determine the distance is not divided in half to take into account the return time of the pulse because the sample program is for the Basic Stamp II, which returns PULSIN values in 2 µs increments. Because the PULSIN command with PicBasic Pro is returning values in increments of 10 µs, the conversion factor will need to be divided by 5, so that we get the correct value based on our 10 µs increment. Taking the PULSIN increment timing difference into account gives us an approximate conversion factor of 15 for inches and 6 for centimeters. Testing with the ranger indicated that the raw value returned by PULSIN when an object was 12 inches away was 180. One hundred and eighty divided by the inch conversion factor of 15 gives us the correct distance of 12 inches.

In order to test the SRF04 sonar ranger, a program will be written to produce audible tones, based on the distance of an object from the device. Compile the sonar-test.bas code listed in **Program 7.5,** and then program the PICmicro MCU 16F84 with the sonar-test.hex file listed in **Program 7.6**. When the PIC is inserted into the 18-pin socket on the main controller board and power is applied, move your hand slowly toward the ranger and notice that the tones produced by the PIC get lower the closer your hand gets to the device. If no tones are produced when power is applied, then check to make sure that none of the connections from the sonar module to the controller board have been mixed up.

PROGRAM 7.5

sonar-test.bas program listing

```
'_____

' Name    : sonar-test.bas
' Compiler : PicBasic Pro - MicroEngineering Labs
' Notes    : Program control of the Devantech SRF04
'           : ultrasonic module. Convert the raw distance
'           : data to a frequency and output to the piezo
'           : element.
'_____


' PortA set as outputs. Pin 1 input.
trisa = %00000010

' PortB set as outputs.
trisb = %00000000


'_____

' initialize variables

trigger      VAR PORTA.0
echo         VAR PORTA.1
piezo        VAR PORTA.3
dist_raw     VAR WORD
dist_inch    VAR WORD
dist_cm      VAR WORD
freq         VAR WORD
conv_inch    CON 15
conv_freq    CON 6
```

SOUND PIEZO,[115,10,50,10]

PROGRAM 7.5

sonar-test.bas program
listing (continued)

start:

main:

```
     gosub sr_sonar



     if freq > 47 then main

     sound piezo,[80 + freq,10]

Goto main

'_____


sr_sonar:

          pulsout trigger,1              ' send a 10us trigger pulse to
                                           the SRF04

          pulsin echo,1,dist_raw        ' start timing the pulse width
                                           on echo pin

          dist_inch = (dist_raw/conv_inch) ' Convert raw data into inches

          freq = (dist_raw/conv_freq)   ' Convert raw data into a
                                           frequency

          pause 10                      ' wait for 10ms before
                                           returning to main

return

end
```

PROGRAM 7.6

sonar-test.hex file
listing

```
:10000000C828A2008417800484138E010C1C8E0063
:1000100023200319C32823200319C3282320C3281E
:10002000A20059200C080D040319C328BD20841315
:100030002208800664001C281D288C0A03198D0FD5
:100040001A288006C32822088E0601308C008D01F4
:100050000000822050E06031D08008C0A03198D0FE7
:100060002828008F002408840022095A208413BD
:100070008F080319C328F03091000E0880389000D3
:10008000F03091030319910003198F030319C3285A
:1000900049285D2003010C1822088E1F22088E08B3
:1000A00003190301900F562880063D2857280000A9
:1000B0004028FF3A84178005C3280D080C04031953
:1000C0008C0A80300C1A8D060C198D068C188D0642
:1000D0000D0D8C0D8D0DC3288F018E00FF308E0706
:1000E00031C8F07031CC32803308D00DF307A20E8
:1000F0006E288D01E83E8C008D09FC30031C83289E
:100100008C07031880288C0764008D0F80280C183A
:1001100089288C1C8D2800008D2808008E00013055
:10012000912894000F080D02031D98280E080C0258
:100130004300318013003190230140503 1DFF3089
:10014000C328910190011030920000D0D900D910D7A
:100150000E0890020F08031C0F0F91020318B72816
:100160000E0890070F0803180F0F910703108C0D4E
:100170008D0D920BA5280C08C3288C098D098C0ABB
:100180000003198D0A0800831303138312640008 0007
:100190008316023085000130860005308312A400EA
:1001A0000830A20073308E000A30322032308E00C8
:1001B0000A303220F4202C088C002D088D008F018D
:1001C0002F308E20031DDA280530A4000830A2004D
:1001D00050302C079E002D080318013E9F001E087A
:1001E0008E000A303220DA2801308C008D01053073
:1001F00084000130102001308C0005308400023072
:100200000001200C08AA000D08AB002A088C002B085E
:100210008D000F308E008F01A120A8000D08A900CD
:100220002A088C002B088D0006308E008F01A1203B
:10023000AC000D08AD000A306C20080063001E29D8
:02400E00F53F7C
:00000001FF
```

Obstacle Avoidance Using the Ultrasonic Range Finder

In the next experiment, the robot will explore its environment and will react to obstacles based on the distance information obtained from the SRF04 sonar module. The robot will normally travel in a forward direction while sonar distance measurements are taken. When it is determined that the robot is within 12 inches of an object, it will reverse, and then alternate between rotating to the left and rotating to the right each time an obstacle is sensed. The distance that the robot travels in reverse and how far it rotates in either direction is determined by the amount of time that the motors are activated. The robot will rotate a further distance to the right than to the left so that it does not get stuck in corners. You can try experimenting with the pause values to change the behavior. When the avoidance maneuver is complete, the robot will continue moving forward. Compile the avoidance.bas code listed in **Program** 7.7, and then program the PICmicro MCU 16F84 with the avoidance.hex file listed in **Program** 7.8. After watching the robot behavior, it is obvious that a better system to track the distance that the robot has traveled or rotated is needed. Later in the chapter, a linear optical shaft encoder will be added to track distance traveled, and to develop a more precise motor control method.

PROGRAM 7.7

avoidance.bas program listing

```
'_____

' Name     : avoidance.bas
' Compiler : PicBasic Pro - MicroEngineering Labs
' Notes    : Obstacle avoidance using the sonar ranger
'_____


' PortA set as outputs. Pin 1 input.
trisa = %00000010

' PortB set as outputs.
trisb = %00000000
```

PROGRAM 7.7

avoidance.bas program
listing (continued)

```
'_____

' initialize variables

trigger              VAR PORTA.0
echo                 VAR PORTA.1
piezo                VAR PORTA.3
enable_right         VAR PORTB.1
forward_right        VAR PORTB.2
reverse_right        VAR PORTB.3
enable_left          VAR PORTB.4
reverse_left         VAR PORTB.5
forward_left         VAR PORTB.6
dist_raw             VAR WORD
dist_inch            VAR WORD
conv_inch            CON 15
turn                 VAR BYTE


low enable_left
low forward_left
low reverse_left

low enable_right
low forward_right
low reverse_right

SOUND PIEZO,[115,10,50,10]
turn = 0

start:

gosub sr_sonar

if dist_inch < 12 then
   turn = turn + 1
   gosub backwards
   if turn.0 = 1 then
      gosub turn_left
   else
```

```
        gosub turn_right
    endif
endif

gosub forward

goto start
'_____

' movement subroutines

forward:

    high enable_left
    high forward_left
    high enable_right
    high forward_right

    pause 50

    low enable_left
    low forward_left
    low enable_right
    low forward_right

return


'_____


turn_left:

    high enable_left
    high forward_left
    high enable_right
    high reverse_right

    pause 300

    low enable_left
```

PROGRAM 7.7

avoidance.bas program
listing (continued)

PROGRAM 7.7

avoidance.bas program
listing (continued)

```
            low forward_left
            low enable_right
            low reverse_right

        return

        '_____

        backwards:

            SOUND PIEZO,[115,5,90,2,80,4,50,10]

            high enable_left
            high reverse_left
            high enable_right
            high reverse_right

            pause 300

            low enable_left
            low reverse_left
            low enable_right
            low reverse_right

        return

        '_____

        turn_right:

            high enable_left
            high reverse_left
            high enable_right
            high forward_right

            pause 600

            low enable_left
```

```
    low reverse_left
    low enable_right
    low forward_right

return

'_____

sr_sonar:

    pulsout trigger,1

    pulsin echo,1,dist_raw
    dist_inch = (dist_raw/conv_inch)
    pause 10

return

end
```

PROGRAM 7.7

avoidance.bas program
listing (continued)

```
:10000000C828A0008417800484138E010C1C8E0065
:1000100023200319C32823200319C3282320C3281E
:10002000A00059200C080D040319C328BD20841317
:100030002008800664001C281D288C0A03198D0FD7
:100040001A288006C32820088E0601308C008D01F6
:100050000000820050E06031D08008C0A03198D0FE9
:10006000282808008F002208840020095A208413C1
:100070008F080319C328F03091000E0880389000D3
:10008000F0309103031991000 3198F030319C3285A
:1000900049285D2003010C1820088E1F20088E08B7
:1000A00003190301900F562880063D2857280000A9
:1000B0004028FF3A84178005C3280D080C04031953
:1000C0008C0A80300C1A8D060C198D068C188D0642
:1000D0000D0D8C0D8D0DC3288F018E00FF308E0706
:1000E000031C8F07031CC32803308D00DF307A20E8
:1000F0006E288D01E83E8C008D09FC30031C83289E
:100100008C07031880288C0764008D0F80280C183A
:1001100089288C1C8D2800008D2808008E00033053
```

PROGRAM 7.8

avoidance.hex file
listing

PROGRAM 7.8

avoidance.hex file
listing (continued)

```
:10012000912894000F080D02031D98280E080C0258
:100130000430031801300319023014050310FF3089
:10014000C328910190011030920000D0D900D910D7A
:100150000E0890020F08031C0F0F91020318B72816
:100160000E0890070F0803180F0F910703108C0D4E
:100170008D0D920BA5280C08C3288C098D098C0ABB
:1001800003198D0A08008313031383126400080007
:100190008316033085000130860083120612831611
:1001A0000612831206138316061383128612831611
:1001B00086128312861083168610831206118316B8
:1001C0000611831286118316861105308312A20050
:1001D0000830A00073308E000A30322032308E009A
:1001E0000A303220A801AD2124088C0025088D009A
:1001F0008F010C308E20031D0529A80A4F216400B1
:10020000281C04292A21052988210721F3280616FC
:100210008316061283120617831606138312861498
:100220008316861083120615831606113230831248
:100230006C20061283160612831206138316061309
:100240008312861083168610831206118316061FF8
:1002500083120800061683160612831206178316E9
:1002600006138312861483168610831286158316E4
:100270008611831201308F002C306D2006128316F8
:100280000612831206138316061383128610831632
:100290008610831286118316861183120800053090
:1002A000A2000830A00073308E00053032205A3092
:1002B0008E000230322050308E0004303220323036
:1002C0008E000A303220061683160612831286161616
:1002D0008316861283128614831686108312861555F
:1002E0008316861183120130 8F002C306D20061288
:1002F0008316061283128612831686128312861044
:100300008316861083128611831686118312080045
:100310000616831606128312861683168612831219
:100320008614831686108312061583160611831200F
:1003300002308F0058306D2006128316061283128B
:100340008612831686128312861083168610831285
:1003500006118316061183120800011308C008D0FE
:10036000053084000130102001308C0005308400FD
:100370000230012000C08A6000D08A70026088C00FA
```

:1003800027088D000F308E008F01A120A4000D08DA
:0C039000A5000A306C2008006300CC2996
:02400E00F53F7C
:00000001FF

PROGRAM 7.8

avoidance.hex file
listing (continued)

Now that radio remote control and sonar obstacle avoidance has been covered, a program will be written to incorporate both. An operator will determine the robot movements, via the remote control. Based on distance measurements taken from the sonar module, the microcontroller will inhibit movement if the robot is in danger of crashing into an obstacle. Since the sonar is mounted to the front of the robot, this will help protect the device. Compile the program called remote-sonar.bas, listed in **Program** 7.9. Program the PIC 16F84 with the corresponding remote-sonar.hex file, listed in **Program** 7.10.

This kind of human/machine interaction is valuable in situations where a robot is operated over large distances (teleoperated). Because the complexity of machines has increased, it is impossible for humans to control all of the small aspects of robotic behavior. With teleoperated robotics, the human gives basic control commands and the robot carries out the tasks all on its own. Another problem with controlling machines over large distances, because of slow radio signals, is that time delays are introduced between the human control commands and the robot's actions. If a human operator makes a mistake, the robot will compensate to avoid a catastrophic failure.

PROGRAM 7.9

remote-sonar.bas
program listing

```
'_____

' Name    : remote-sonar.bas
' Compiler : PicBasic Pro - MicroEngineering Labs
' Notes   : Remote control with sonar avoidance.
'_____


' PortA set as outputs. Pin 1 input.
trisa = %00000010
```

PROGRAM 7.9

remote-sonar.bas
program listing
(continued)

```
' PortB set as outputs. pin 0 input.
trisb = %00000001

'_____

' initialize variables

include "modedefs.bas"

rx_baud          CON N2400
rxmit            VAR PORTB.0
enable_right     VAR PORTB.1
forward_right    VAR PORTB.2
reverse_right    VAR PORTB.3
enable_left      VAR PORTB.4
reverse_left     VAR PORTB.5
forward_left     VAR PORTB.6
trigger          VAR PORTA.0
echo             VAR PORTA.1
piezo            VAR PORTA.3
control          VAR BYTE
temp             VAR BYTE
dist_raw         VAR WORD
dist_inch        VAR WORD
conv_inch        CON 15
low enable_left
low forward_left
low reverse_left

low enable_right
low forward_right
low reverse_right

SOUND PIEZO,[115,10,50,10]

start:

serin rxmit,rx_baud,["Z"],control
```

PROGRAM 7.9

remote-sonar.bas
program listing
(continued)

```
if control = "A" then
   gosub sr_sonar
   if dist_inch < 8 then start
   gosub forward
endif

if control = "B" then
   gosub backwards
endif

if control = "C" then
   gosub turn_left
endif

if control = "D" then
   gosub turn_right
endif

if control = "E" then
   sound piezo,[115,10,50,10]
endif

if control = "F" then
   low enable_left
   low forward_left
   low reverse_left
   low enable_right
   low forward_right
   low reverse_right
endif

goto start
'_____

' movement subroutines

forward:
```

PROGRAM 7.9

remote-sonar.bas
program listing
(continued)

```
high enable_left
high forward_left
high enable_right
high forward_right

pause 500

low enable_left
low forward_left
low enable_right
low forward_right

return

'_____


turn_left:

    high enable_left
    high forward_left
    high enable_right
    high reverse_right

return

'_____


backwards:

    high enable_left
    high reverse_left
    high enable_right
    high reverse_right

return

'_____
```

turn_right:

```
high enable_left
high reverse_left
high enable_right
high forward_right
```

return

'_____

sr_sonar:

```
pulsout trigger,1
pulsin echo,1,dist_raw
dist_inch = (dist_raw/conv_inch)
pause 10
```

return

end

PROGRAM 7.9

remote-sonar.bas
program listing
(continued)

```
:100000000629A000841780048413BE010C1C8E0026
:10001000232003190129232003190129232001296 1
:10002000A0008F200C080D0403190129FB20841364
:1000300020088006640001C281D288C0A03198D0FD7
:100040001A288006012920088E0601308C008D01B7
:1000500000082005 0E06031D08008C0A03198D0FE9
:1000600028280800640042200318322 84D20083058
:100070008F004E2042208E0C3D288F0B39284E20B9
:100080000E080800220884002008841780048413C6
:100090000000051F192006FF3E08001F171F0D063917
:1000A0008C005C208D008C0A5C201F1FB0281F1361
:1000B0008C000230C720B02800308A000C0882076C
:1000C0000001347534033415340034 3C340C34D934E1
:1000D0008F0022088400200990208413 8F080319C0
:1000E0000129F03091000E0880389000F030910323
:1000F000031991000319 8F03031901297F28932005
```

PROGRAM 7.10

remote-sonar.hex file
listing

PROGRAM 7.10

remote-sonar.hex file
listing (continued)

```
:1001000003010C1820088E1F20088E080319030114
:10011000900F8C28800673288D2800007628FF3ADF
:100120008417800501290D080C0403198C0A8030FE
:100130000C1A8D060C198D068C188D060D0D8C0D64
:100140008D0D01298F018E00FF308E07031C8F0754
:10015000031C012903308D00DF30B020A4288D015D
:10016000E83E8C008D09FC30031CB9288C0703186D
:10017000B6288C0764008D0FB6280C18BF288C1C7D
:10018000C3280000C328080003108D0C8C0CFF3E10
:100190000318C4280C0801298E000430CF289400CD
:1001A0000F080D02031DD6280E080C020430031898
:1001B000013003190230140503 1DFF30012991019C
:1001C0009001103092000D0D900D910D0E089002CF
:1001D0000F08031C0F0F91020318F5280E08900753
:1001E0000F08 03180F0F910703108C0D8D0D920B44
:1001F000E3280C0801298C098D098C0A03198D0A42
:10020000080083130313831264000800831602306E
:10021000850001308600831206128316061283 12AF
:10022000061383160613831286128316 8612831210
:10023000861083168610831206118316 0611831208
:100240008611831686110530 8312A2000830A000A3
:1002500073308E000A30682032308E000A306820F9
:100260000630A2000130A00004309F0032205A3C2A
:1002700031D36293220A80064002808413C031DD4
:100280004C29E52124088C0025088D008F010830B9
:10029000CC20031D30298D2164002808423C031D19
:1002A0005229C32164002808433C031D5829B22168
:1002B00064002808443C031D5E29D42164002808FA
:1002C000453C031D6F290530A2000830A0007330A3
:1002D0008E000A30682032308E000A3068206400B8
:1002E0002808463C031D8C2906128316061283 1229
:1002F00006138316061383128612831686128312 40
:10030000861083168610831206118316 0611831237
:10031000861183168611831230290616831606125B
:1003200083120617831606138312861483168 6100B
:100330008312061583160611831201308F00F430E4
:10034000A32006128316061283120613831606 13C1
:100350008312861083168610831206118316 0611E7
```

```
:100360008312080006168316061283120617B316D8
:1003700006138312861483168610831286158316D3D
:1003800086118312080006168316061283128616B3B
:1003900083168612831286148316861083128615B9E
:1003A000831686118312080006168316061283121E
:1003B0008616831686128312861483168610831267D
:1003C0000061583160611831208000130BC008D017A
:1003D000053084000130102001308C00053084008D
:1003E000023001200C08A6000D08A70026088C008A
:1003F00027088D000F308E008F01DF20A4000D082C
:0C040000A5000A30A22008006300042AB6
:02400E00F53F7C
:00000001FF
```

PROGRAM 7.10

remote-sonar.hex file
listing (continued)

Distance Measurement Using an Optical Shaft Encoder

A shaft encoder is a sensor that measures the position or velocity of a shaft. Shaft encoders are generally inexpensive devices that are most often mounted on the output shaft of a drive motor or on the axle. The signal that is produced by this sensor can be either a code that corresponds to a particular position of the shaft (called absolute encoders), or it may be a pulse train. Shaft encoders that produce a pulse train are called incremental encoders. The encoder is typically a disk that has numerous holes or slots along its outside edge. An infrared LED is placed on one side of the disk and an infrared-sensitive phototransistor is positioned directly opposite the LED. As the shaft rotates, the holes pass the light intermittently and the state of the phototransistor output changes from high to low or vise versa, producing a pulse train. The rate at which the pulses are produced corresponds to the rate at which the shaft turns. By using a microprocessor to count the pulses, the robot can determine how far its wheels have rotated. The combination of an infrared LED emitter and a phototransistor, packaged

FIGURE 7.27

Optical encoder and
photointerrupter
attached to motor
shaft.

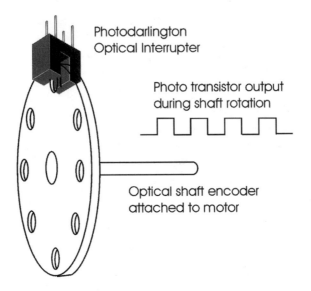

Photodarlington
Optical Interrupter

Photo transistor output
during shaft rotation

Optical shaft encoder
attached to motor

for the purpose of being mounted on either side of a shaft encoder's disk, is called a photointerrupter, as shown in **Figure 7.27**.

The device that we will use to track the distance that Turtletron's motor has traveled is a photodarlington optical interrupter switch with part number H22B1, shown in **Figure 7.28**. The H22B1 consists of a gallium arsenide infrared LED, coupled with a silicon photodarlington in a plastic housing. The packaging system is designed to optimize the mechanical resolution, coupling efficiency, ambient light rejection, cost, and reliability. The gap in the housing provides a means of interrupting the signal with an opaque material, switching the output from an "ON" to an "OFF" state. The interrupter will be mounted on a small circuit board and placed around the encoder disk that was included with the motor kits.

PACKAGE DIMENSIONS

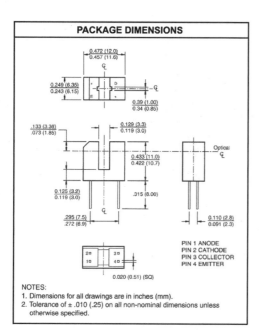

0.472 (12.0)
0.457 (11.6)

0.249 (6.35)
0.243 (6.15)

0.39 (1.00)
0.34 (0.85)

.133 (3.38)
.073 (1.85)

0.129 (3.3)
0.119 (3.0)

Optical

0.433 (11.0)
0.422 (10.7)

0.125 (3.2)
0.119 (3.0)

.315 (8.00)

.295 (7.5)
.272 (6.9)

0.110 (2.8)
0.091 (2.3)

PIN 1 ANODE
PIN 2 CATHODE
PIN 3 COLLECTOR
PIN 4 EMITTER

0.020 (0.51) (SQ)

NOTES:
1. Dimensions for all drawings are in inches (mm).
2. Tolerance of ± .010 (.25) on all non-nominal dimensions unless otherwise specified.

OPTICAL INTERRUPTER SWITCH PHOTODARLINGTON

SCHEMATIC

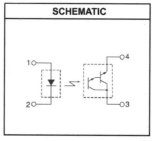

FIGURE 7.28

H22B1 interrupter switch package dimensions and schematic.

Fabricating the Shaft Encoder

Start by locating the plastic motor mount with six evenly spaced holes along its outside edge. It is shown in **Figure 7.29** and will function as the encoder disk. Mount it to the end of the robot's left motor shaft, opposite to the wheel, using the washer and nut that were supplied. This small disk only has six holes and six opaque areas, giving us a rotational accuracy of 30 degrees per state change, which is sufficient for a small robot like Turtletron. If the microcontroller is monitoring the interrupter switch and counts 12 state changes, from high to low or vice versa, then the wheel has made one full rotation. The parts needed to construct the shaft encoder are listed in **Table 7.5**.

FIGURE 7.29

Encoder disk.

Part	Quantity	Description
H22B1 Optical interrupter switch	1	Photodarlington transistor
R1, R2	2	1 KΩ 1/4-watt resistor
R3	1	470 KΩ 1/4-watt resistor
D1	1	Red light-emitting diode
Three-strand ribbon wire	7 inches	Connector wire
header	1	4-post male connector— 2.5-mm spacing
Hot glue	—	Hot glue and gun

TABLE 7.5

Optical Encoder Parts List

The schematic to interface the interrupter switch to the PIC 16F84 is shown in **Figure 7.30**. The circuit operates by using the transistor as a switch. Cut a piece of perfboard to a size of 1-1/2 inches × 3/4 of an inch. Create an aluminum mounting bracket by following the cutting, drilling, and bending instructions in **Figure 7.31**. Use 1/16-inch thick aluminum to construct the mounting bracket. When it is complete, position the 3/4-inch side of the mount on the left side of the perfboard and mark the mounting hole. Drill the hole with a 5/32-inch bit. Attach the mount to the perfboard with a 6/32-inch × 1/2-inch machine screw and lock-

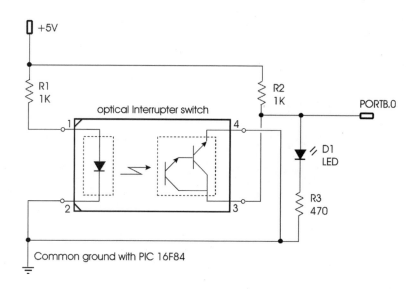

FIGURE 7.30

Schematic to interface the H22B1 interrupter switch to PIC 16F84.

ing nut. Mount the interrupter to the upper left side of the board and secure it in place with hot glue. Because the circuit is so simple, use point to point wiring to solder all of the parts together. Mount the LED with the leads bent on a 90-degree angle so that it is pointing upwards. The LED will indicate when the infrared light beam has been interrupted by the opaque parts of the disk. Cut a piece of 3-strand ribbon wire to a length of 7 inches. Solder the end of one wire to the 5-volt connection point, another wire to terminal 3 of the H22B1, and the last wire to the common ground point. The finished interface board is shown in **Figure 7.32.**

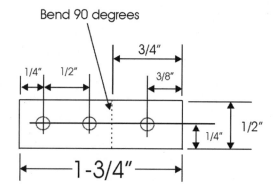

FIGURE 7.31

Encoder board mounting bracket.

FIGURE 7.32

Finished encoder
interface board.

Place the interface board on the bottom of the robot so that the
encoder disk is positioned in between the H22B1 interrupter. Make
sure that the disk does not touch either side of the interrupter and
can rotate freely when the tire is turned by hand. Mark the position of the two mounting holes on the robot base. Drill the two
holes with a 5/32-inch bit, and then secure the interface board
with two 6/32-inch × 1/2-inch machine screws and locking nuts.
Rotate the tire by hand. Make any necessary position adjustments
of the interface board if the disk is not centered. A close-up of the
encoder disk and interface board is shown in **Figure 7.33**. To get
an idea of the overall positioning, refer to **Figure 7.34**. Drill a hole
in the robot base and feed the 3-strand connector wire through.
Disconnect the RF receiver module from the main controller board.
The interrupter circuit will use the connector that the RF module
was plugged into. Follow the wiring diagram in **Figure 7.35** to
connect the interface board to the main controller.

FIGURE 7.33

Optical encoder disk centered between the interrupter.

FIGURE 7.34

Interrupter interface board mounted to robot base.

FIGURE 7.35

Wiring diagram to connect interface board to main controller.

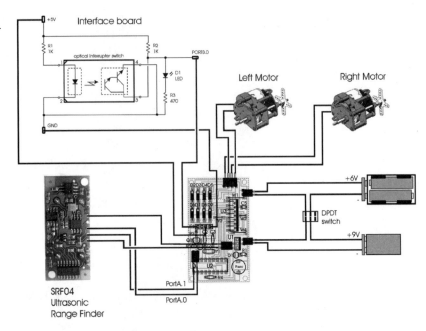

When the photointerrupter is connected to the main controller, take the PIC microcontroller out of the 18-pin socket and turn on the power. Slowly rotate the tire by hand. The LED will be on when an opaque section of the encoder disk is between the optical interrupter. When it comes across a hole in the disk, the LED will be off. The next program will test the connection of the device to the PIC 16F84 microcontroller. Compile encode-test.bas, listed in **Program 7.11**, and then program the PIC 16F84 with the encode-test.hex file listed in **Program 7.12**. Place the PIC in the 18-pin socket and then turn on the power. When you rotate the tire and encoder disk by hand, the PIC will produce a sound each time a hole is encountered. The program stays in a tight loop until the transistor changes state again; otherwise the PIC would continuously produce the tone sequence while the disk was on the same hole. This method will be used when counting the number of times the transistor switches from one state to another, or an event is being triggered. If a counter is being incremented, this method ensures that only one count will occur during a state transition.

```
'_____
' Name     : encode-test.bas
' Compiler : PicBasic Pro - MicroEngineering Labs
' Notes    : Program to test the optical interrupter
'            : photodarlington switch
'_____


' PortA set as outputs.
trisa = %00000000

' PortB set as outputs. pin 0 input.
trisb = %00000001


'_____

' initialize variables

switch          VAR PORTB.0
enable_right    VAR PORTB.1
forward_right   VAR PORTB.2
reverse_right   VAR PORTB.3
enable_left     VAR PORTB.4
reverse_left    VAR PORTB.5
forward_left    VAR PORTB.6
piezo           VAR PORTA.3
control         VAR BYTE
temp            VAR BYTE

low enable_left
low forward_left
low reverse_left

low enable_right
low forward_right
low reverse_right

SOUND piezo,[115,10,50,10]

start:
```

PROGRAM 7.11

encode-test.bas
program listing
(continued)

```
If switch = 0 then
    SOUND piezo,[80,5,110,5,50,10,120,2]
    while switch = 0
    wend
endif

goto start
```

PROGRAM 7.12

encode-test.hex file
listing

```
:100000003F288F0022088400200928208413BF08AD
:1000100003193A28F03091000E0880389000F03033
:100020009103031991000319BF0303193A28182823
:100030002B2003010C1820088E1F20088E0803199E
:10004000301900F252880060C28262800000F2881
:100050008417800053A280D080C0403198C0A803097
:100060000C1A8D060C198D068C188D060D0D8C0D35
:100070008D0D3A2883130313831264000B0083163E
:100080008501013086008312061283160612831240
:100090000061383160613831286128316861 28312A2
:1000A0008610831686108312061183160611831 29A
:1000B0008611831686110530831 2A2000830A00035
:1000C00073308E000A30012032308E000A30012059
:1000D000640006188328053D A2000830A0005030C4
:1000E0008E00053001206E308E0005300120323048
:1000F0008E000A30012078308E000230012064002A
:080100006188328 7F286828F7
:02400E00F53F7C
:00000001FF
```

Room Mapping Using the Shaft Encoder and Ultrasonic Range Finder

The robot now has the ability to keep track of how far the left wheel has traveled using the incremental shaft encoder. This will be necessary when the robot is mapping an area before it starts to move. In previous programs where the robot used the sonar ranger, it avoided obstacles in a reactionary way because it did not have an internal representation of the outside world. It wandered around the room until distance readings from the sonar module alerted the robot that an evasive maneuver was needed to avoid crashing into an obstacle.

To improve this situation, the robot will need to create a rudimentary map of the area surrounding its current position. A robot's ability to create an internal representation of the external world can be thought of as the first measure of machine intelligence, and is a necessary evolutionary step to self awareness and consciousness. The final program in this chapter will take advantage of the optical shaft encoder and the ultrasonic range finder to give the robot the ability to map the area around itself and store the results internally. Based on this information, the robot can then make an intelligent decision about where to move.

This is accomplished by having the robot take a series of distance measurements in a 180-degree arc to the front and sides of its current location. From where the robot is facing, it will rotate 90 degrees to the left and then start taking distance measurements as it rotates back in the opposite direction for 180 degrees. The distance measurements are stored in a one-dimensional array called *position*, made up of 12 elements. To make sure that the robot is consistently moving the same distance for each sonar measurement taken, the output from the optical encoder circuit is used. The motor control algorithm works by first reading the current state of the sensor. The initial state of the sensor doesn't matter; we are concerned with when the sensor changes from its current state, indicating that the wheel has moved 1/12 of a complete rotation. Using this method makes motor control and wheel tracking uncomplicated. The program takes the current state of the sensor and stores it in a variable. The motor is then moved by a very small amount, and the stored sensor state is then compared to the current state. If the two states are the same, then the motor is moved again by a small amount. This continues until the sensor has changed from its original state, at which time the motor is stopped and the next sonar distance reading is taken. This indicates that the motor has moved the wheel by 1/12 of a complete rotation.

When all of the sonar distance measurements have been taken, a sorting algorithm determines which position contains the distance measurement with the highest value. The robot is then rotated back to the position with the greatest amount of free space, and then moves forward to map out the surrounding area. If an obstacle is encountered while moving forward, the robot backs up and makes another map to determine the best route to take. Compile sonar-map.bas, listed in **Program 7.13**, and then program the PIC 16F84 with the corresponding sonar-map.hex file, listed in **Program 7.14**.

I find this final experiment to be a lot of fun because of the speed at which the robot scans the area while making maps, and how fast it can travel through a room. It is very surprising to see how well the robot can maneuver through rooms and consistently pick the areas with the most free space.

To develop robotic room mapping further, write a program that stores the distance readings in a two-dimensional array. This way the robot would be able to quickly backtrack without having to take sonar readings for an area that it has already explored.

PROGRAM 7.13

sonar-map.bas program listing

```
'_____
' Name    : sonar-map.bas
' Compiler : PicBasic Pro - MicroEngineering Labs
' Notes    : Room mapping using the sonar ranger and
'           : incremental shaft encoder.
'_____


' PortA set as outputs. Pin 1 input.
trisa = %00000010


' PortB set as outputs. pin 0 input.
```

```
trisb = %00000001
```

PROGRAM 7.13

sonar-map.bas program
listing (continued)

```
'_____

' initialize variables

trigger          VAR PORTA.0
echo             VAR PORTA.1
piezo            VAR PORTA.3
switch           VAR PORTB.0
enable_right     VAR PORTB.1
forward_right    VAR PORTB.2
reverse_right    VAR PORTB.3
enable_left      VAR PORTB.4
reverse_left     VAR PORTB.5
forward_left     VAR PORTB.6
dist_raw         VAR WORD
dist_inch        VAR WORD
conv_inch        CON 15
I                VAR BYTE
temp             VAR BYTE
state            VAR BYTE
best_pos         VAR BYTE
most_space       VAR BYTE
position         VAR WORD[12]

low enable_left
low forward_left
low reverse_left

low enable_right
low forward_right
low reverse_right

SOUND PIEZO,[115,10,50,10]

start:

' rotate robot to the left
```

PROGRAM 7.13

sonar-map.bas program
listing (continued)

```
For I = 1 to 5
    state = switch
    while switch = state
        gosub turn_left
    wend
Next I

position[11] = 0

' take 11 distance measurements and store the
' results in the distance[11] array

For I = 0 to 10
    gosub sr_sonar
    position[I] = dist_inch
    state = switch
    while switch = state
        gosub turn_right
    wend
Next I

' sort the distance array to find the location
' with the most free space

best_pos = 11

For I = 0 to 10
    If position[I] >= position[best_pos] then
        best_pos = I
    Endif
Next I

most_space = 11 - best_pos

' rotate the robot so that it is pointing towards
' the area with the most free space
```

PROGRAM 7.13

sonar-map.bas program
listing (continued)

```
For I = 1 to most_space
   state = switch
   while switch = state
      gosub turn_left
   wend
Next I


' Move the robot forward into the area that was
' determined to be the most free of obstacles.
' Check for any obstacles while moving forward.
' Move in reverse and then scan for area with
' most space if an obstacle was encountered.

For I = 1 to 24
   gosub sr_sonar
   If dist_inch < 8 then
      SOUND PIEZO,[115,5,90,2,80,4,50,10]
      For temp = 1 to 6
         state = switch
         while switch = state
         gosub backwards
         wend
      Next temp
      goto start
   Endif

   state = switch
   while switch = state
      gosub forward
   wend

Next I

goto start

end

'_____
```

PROGRAM 7.13

sonar-map.bas program
listing (continued)

```
' movement subroutines

forward:

    high enable_left
    high forward_left
    high enable_right
    high forward_right

    pause 20

    low enable_left
    low forward_left
    low enable_right
    low forward_right

    pause 20

return

'_____


turn_left:

    high enable_left
    high forward_left
    high enable_right
    high reverse_right

    pause 5

    low enable_left
    low forward_left
    low enable_right
    low reverse_right

    pause 5
```

return

'————————————————————————————————————

backwards:

```
    high enable_left
    high reverse_left
    high enable_right
    high reverse_right

    pause 20

    low enable_left
    low reverse_left
    low enable_right
    low reverse_right

    pause 10

return
```

'————————————————————————————————————

turn_right:

```
    high enable_left
    high reverse_left
    high enable_right
    high forward_right

    pause 5

    low enable_left
    low reverse_left
    low enable_right
    low forward_right
```

PROGRAM 7.13

sonar-map.bas program
listing (continued)

PROGRAM 7.13

sonar-map.bas program
listing (continued)

```
pause 5

return

'_____

sr_sonar:

    pulsout trigger,1

    pulsin echo,1,dist_raw
    dist_inch = (dist_raw/conv_inch)
    pause 2

return

end
```

PROGRAM 7.14

sonar-map.hex file
listing

```
:10000000CB28A4008417800484138E010C1C8E005E
:1000100023200319C62823200319C6282320C62815
:10002000A40059200C080D040319C628C02084130D
:100030002408800664001C281D288C0A03198D0FD3
:100040001A288006C62824088E0601308C008D01EF
:10005000000824050E06031D08008C0A03198D0FE5
:10006000282808008F002608840024095A208413B9
:100070008F080319C628F03091000E0880389000D0
:10008000F03091030319910003198F030319C62857
:1000900049285D2003010C1824088E1F24088E08AF
:1000A00003190301900F562880063D2857280000A9
:1000B0004028FF3A84178005C6280D080C04031950
:1000C0008C0A80300C1A8D060C198D068C188D0642
:1000D0000D0D8C0D8D0DC6288F018E00FF308E0703
:1000E000031C8F07031CC62803308D00DF307A20E5
:1000F0006E288D01E83E8C008D09FC30031C83289E
:100100008C07031880288C0764008D0F80280C183A
:1001100089288C1C8D2800008D2808008E00033053
:10012000094288E000430942894000F080D02031DBB
:100130009B280E080C02043003180130031902300A
```

:100140001405031DFF30C62891019001103092064
:100150000D0D900D910D0E0890020F08031C0F0F4E
:1001600091020318BA280E0890070F0803180F0F02
:100170000910703108C0D8D0D920BA2880C08C62832
:100180008C098D098C0A03198D0A08008313031347
:100190008312640008008316023085000130860057
:1001A000831206128316061283120613831606 1391
:1001B0008312861283168612831286108316861087
:1001C000831206118316061183128611831686 1177
:1001D00005308312A6000830A40073308E000A3068
:1001E000322032308E000A3032200130C5006400E7
:1001F0000630450203180A29003006180130C700EE
:100200000640047080618013C031D0829DB2100296A
:10021000C50FF728BE01BF01C50164000B304502C0
:1002200003182A294A220310450D283E840040085D
:100230008000840A4108800000300618013 0C700A1
:100240000640047080618013C031D2829252220299F
:10025000C50F0D290B30C400C50164000B304502E9
:1002600003185229 0310450D283E840000089E0003
:10027000840A00089F000310440D283E84000008F3
:10028000A000840A0008A1001E088C001F088D0031
:1002900021088F0020089120031D50294508C40023
:1002A000C50F2D2944080B3CC6000130C500640071
:1002B00045084602031C6A29003006180130C700B1
:1002C0000640047080618013C031D6829DB216029EA
:1002D000C50F57290130C5006400193045020318C5
:1002E000B3294A2240088C0041088D008F01083054
:1002F0008E20031DA5290530A6000830A400733008
:100300008E00053032205A308E00023032205030BC
:100310008E000430322032308E000A30322001301C
:10032000C8006400073048020318A42900300618EA
:10033000130C700640047080618013C031DA229CC
:100340000229A29C80F9129F5280030061801309B
:10035000C700640047080618013C031DB129B621F7
:10036000A929C50F6C29F5286300B4290616831640
:100370006128312061783160613831286148316 39
:1003800086108312061583160611143083126C2012
:1003900006128316061283120613831606138312 9F

PROGRAM 7.14

sonar-map.hex file
listing (continued)

PROGRAM 7.14

sonar-map.hex file
listing (continued)

```
:1003A00086108316861083120611831606111430E8
:1003B00083126C200800061683160612831206179 5
:1003C0008316061383128614831686108312861 5ED
:1003D0008316861105308312 6C20061283160612CE
:1003E00083120613831606138312861083168610 53
:1003F00083128611831686110530831 26C20080043
:100400000616831606128312861683168612831228
:1004100086148316861083128615831686111430 6F
:100420008312 6C200612831606128312861283161C
:1004300086128312861083168610831286118316 05
:1004400086110A3083126C20080061683160612E 5
:10045000831286168316861283128614831686100DC
:100460008312061583160611053083126C200612BE
:100470008316061283128612831686128312861042
:1004800083168610831206118316061105308312 17
:100490006C20080001308C008D0105308400013093
:1004A000102001308C0005308400023001200C083F
:1004B000C2000D08C30042088C0043088D000F30B5
:1004C0008E008F01A420C0000D08C10002306C20F6
:0604D00008006300692A28
:02400E00F53F7C
:00000001FF
```

8

Taking It Further

After building some or all of the biologically inspired robots in this book, you may have thought of a number of ways to improve or enhance each of the projects. You may have even come up with ideas for completely new robots. If that is the case, then *Amphibionics* has achieved its goal. Listed below are some ideas to take each of the robot projects further.

Frogbotic

1. Add an infrared or ultrasonic range finder to the robot so that it can avoid obstacles before leaping.

2. Add a servo to the front legs so that they can be turned to the left or right. This will make navigation control much easier when combined with the timed release of the back legs.

3. Waterproof the frog by creating a latex outer skin. Rubber latex can be applied to a mold with a paintbrush, and is available at most model hobby shops. It can be built up in layers until the required thickness is achieved.

Serpentronic

1. Create a scaled surface for the underside of the robot that will allow the snake to slide forward, but produce friction in the opposite direction, much like the skin of a real snake. The skin could be fabricated out of very thin sheets of aluminum, overlapping by 1/8 of an inch.

2. Interface a model airplane transmitter and receiver system for human control of the robot. The use of a long-range remote control system will allow the robot to be guided to exact remote locations. Because the robot snake has a low profile and stealthy nature, it has many uses such as espionage applications, military reconnaissance, safe land mine search, and removal, along with locating survivors in disaster areas.

3. Interface various environmental and weather sensors to monitor remote, rough terrain areas accessible only to small animals, such as a snake. Sensors that can measure temperature and humidity can be added so that readings can be taken at different locations and the information radioed back to a main computer or stored in the robot's memory, to be retrieved at a later date.

4. Interface a global positioning system (GPS) module to the PicMicro MCU, and have the robot move from one defined area to another.

Crocobot

1. Include an obstacle avoidance sensor so that the robot can operate autonomously. Try using a method other than infrared or ultrasonic detection, like a simple whisker switch.

2. Add a gripper so that the robot can pick up objects via the remote control. Program the microcontroller so that when a

push-button command is received from the transmitter, the control stick will then be used to operate the gripper.

3. Install a miniature video/audio camera and transmitter for remote visual operation.

4. Incorporate a digital compass or gyroscope into the control system so that the robot can keep its bearing when it is commanded to walk in a straight line.

Turtletron

1. Add a line-following circuit to the underside of the robot consisting of two sets of light-emitting diodes and phototransistors. The robot can be programmed to follow a predetermined white line that has been placed on the floor. This type of navigation is used in some factories. The reflective tape method is preferred, so that the track can easily be changed.

2. Use rechargeable batteries, and then add a battery charger station so that the robot can recharge its batteries when they run low. It could use line-following capability to find its recharging station.

3. Install a small vacuum system on the bottom of the robot. Use the information from the shaft encoder sensor, and program the robot to start moving in a spiral pattern from the center of the room outward. When the ultrasonic sensor indicates that it is near a wall, program the robot to navigate around the edges of the room and under the furniture.

4. Add a light sensitive resistor to the front of the robot, and interface it to the microcontroller. Have the robot search for the brightest areas of the room or the darkest. If solar panels were added to recharge the batteries, this sort of behavior would be desirable.

Bibliography

Anita M. Flynn, Joseph L. Jones, *Mobile Robots, Inspiration to Implementation*, A K Peters, Massachusetts, 1993, ISBN 1-56881-011-3

H.R. Everett, *Sensors for Mobile Robots*, A K Peters, Massachusetts, 1995, ISBN 1-56881-048-2

Karl Williams, *Insectronics—Build Your Own Walking Robot*, McGraw-Hill, New York, 2003, ISBN 0-07-141241-7

Rodney Brooks, *Flesh and Machines*, Random House, New York, 2002, ISBN 0-375-42079-7

Ed Rietman, *Experiments in Artificial Neural Networks*, 1988, TAB BOOKS Inc, PA, ISBN 0-8306-0237-2

Gordon Mccomb, *The Robot Builder's Bonaza*, McGraw-Hill, New York, 1987, ISBN 0-8306-2800-2

John Iovine, *Robots, Androids, and Animatrons*, McGraw-Hill, New York, 2002, 1998, ISBN 0-07-137683-6

Geoff Simons, Robots, *The Quest For Living Machines*, Sterling Publishing, New York, 1992, ISBN 0-304-34414-1

Steven Levy, *Artificial Life*, Random House, New York, 1992, ISBN 0-679-74389-8

Daniel Crevier, *AI—The Tumultuous History of the Search for Artificial Intelligence*, HarperCollins, New York, 1993, ISBN 0-465-00104-1

Index

Note: Boldface numbers indicate illustrations.

@ command, 32
ADCIN, 32, 257
aluminum stock, 12–15, **12, 13, 14**
analog to digital converters, 25, 30, 256–257
antenna, Turtletron, 295–296, **296**
arithmetic logic unit, 25
artificial intelligence, 273–275
ASM...ENDASM, 32
assembler, 38
avoidance.bas/avoidance.hex, Turtletron, 313–319

band saw, 1, **2**
BAS extension on source files, 36
BASIC, 29, 35
BASIC Stamps, 29, 41
battery pack holder
    Crocobot, 203–208, **208**
    Frogbotic, 92, **93, 94**
    Turtletron, 296–297
    Serpentronic, 128–129, **128, 129, 130,** 139, **139,** 142, **143**
battery power supply
    Crocobot, 226
    Frogbotic, 94–95, 101, **102**
    Serpentronic, 144–145, 158–164
BRANCH, 32

BRANCHL, 32
BUFFERs setting, 31
BUTTON, 32

calibration of infrared sensor board, 154–155, **154**, **155**, 178–179
CALL, 32
capacitors, 29
central processing unit (CPU), 25, 26
CLEAR, 32
CLEARWDT, 32
clock speed, PIC 16F84 MCU and 26–27
code file, 38
Code Protect setting, EPIC Programmer and, 43
command/statements listing for PicBasic Pro Compiler, 32–35
compiler (*See also* PicBasic Pro Compiler), 28
CONFIG.SYS file, 31
continuous rotation modification for servo motors, 55–66, **57–66**
controller board
    Crocobot, 216–226, **216**, **222**, **223**, **224**
    Frogbotic, 94–98, **95**, **101**, **102**
    Serpentronic, 144–148, **144**
    testing, 44–45
    Turtletron, 283–286, **284**
controlling modified servo motors, 66–68
copper boards for PCB, 18–20, **19**
COUNT, 32
Crocobot, 191–269
    analog to digital converters in, 256–257
    battery pack holder for, 203–208, **208**
    battery power supply for, 226
    body covers and tail section for, 202–208, **203–208**
    chassis construction for, 199–202, **199–202**
    component connection and assembly steps in, 226–228, **227**, **228**
    control stick for, **231**, 257–269
    controller board for, 216–226, **216**, **222**, **223**, **224**
    crocobot-switch.bas/crocobot-switch.hex for, 239–241
    crocodilian biology and, 191–193, **192**
    gearbox assembly for, 195–198, **196**, **197**, **198**
    L298 dual full-bridge driver in, 218–221, **219**
    leg assembly in, 213–215, **214**
    leg construction in, 211–212, **212**, **213**
    limit switch wiring in, 209, **209**, **210**
    mechanical construction of, 194–198
    modifications and customizations for, 346–347
    motion programming and control in, 244–269

Crocobot (*continued*)
    motor output shafts for legs in, 211–212, **211**
    motor-test.bas/motor-test.hex for, 242–244
    motors for, 193
    overview of, 193–215
    parts list for, 195, 217–218, 233–234
    PIC 16C71 in, 232–234, **232**
    power switch wiring for, 215, **215**
    programming for, 239–269
    radio transmitter and receiver modules for, 220–221, **221**, 224–239,
        225, 226, 238, 239
    receive-test.bas/receive-test.hex for, 252–254
    remote control printed circuit board, 234–239, **235, 236, 237**
    remote control transmitter for, 228–239, **229–231**
    remote controller for, 193–194, **194**
    rx-remote.bas/rx-remote.hex in, 257, 258–265
    serial data link in, 251–269
    transmit-test.bas/transmit-test.hex in, 254–256
    tx-remote.bas/tx-remote.hex in, 257, 265–269
    walk-routines.bas/walk-routines.hex for, 245–251
    wiring for, 201, **201, 238**
crocobot-switch.bas/crocobot-switch.hex for, 239–241

DATA, 32
DEBUG, 32
debugging, 30
DEBUGIN, 32
Devantech SRF04 ultrasonic range finder, 286–298, **286**
developing PCB, 18–20, **19**
differential drive system, Turtletron, 283
digital multimeter, 10–11, **10**, 154–155, **154**
digital-to-analog converters, 25
DISABLE, 32
distance measurement using optical shaft encoder in, 325–344, **326,
    327**
DOS EDIT, 35
drilling and parts placement on PCB, 21–22, **22, 23**
drills and drill presses, 3–5, **4, 5**
DTMFOUT, 32

EEPROM command, 32
ENABLE, 32, 33
encode-test.bas/encode-test.hex, Turtletron, 332–334
END FOR...NEXT, 33
endmill, 5, **5**

EPIC Programmer, 40–43, **41**
   Code Protect setting in, 43
   DOS users, 41–42
   graphical user interface of, 43, **43**
   opening file in, 42–43
   plugging in, 41–42
   programming process in, 43
   Windows users, 41
epoxy, **9**, 10
erasing and reprogramming PICmicros, 30
etching the PCB, 20–21
exposing PCB, 19–20

fasteners, 14, **15**
ferric chloride for PCB, 18, **19**, 20–21
file, **9**
FILES setting, 31
flash memory, 26, 30
FOR...NEXT, 33
FREQOUT, 33
friction pads, Serpentronic, 163, **164**
frog-test.asm, 38, 39
frog-test.bas, 36–38, 39, 103, 104–106
frog-test.hex, 38, 39–40, 104, 106–107
Frogbotic, 51–116, **103**
   battery pack holder for, 92, **93, 94**
   battery power supply for, 94–95, 101, **102**
   component connections, 100–103, **101**
   controller board for, 94–98, **95**, 101, **102**
   controlling the modified servo in, 66–68
   cutting and bending guides for, 69, **69**
   drilling guide for, **70**
   feet, 76, **76**, 77
   frog and toad biology and, 51–52, **52**
   frog-test.bas/frog-test.hex for, 103, 104–107
   frogbotic.bas/frogbotic.hex for, 111–115
   front leg construction and attachment in, 90, **90**
   jumping motion, programming and controlling, 110–113, **116**
   leg assembly for, 77–82, **78–81**
   leg attachment to body for, 82–83, **82, 83**
   leg pieces for, 74, **75**
   leg position sensors in, 91, **91**
   leg stops for, 74, **74**, 76, **76**
   limit switch wiring in, 91–94, **92, 93**
   limit-switch.bas/limit-switch.hex for, 107–110

Frogbotic (*continued*)
mechanical construction of, 68–77
modifications and customizations for, 345
modifying servos for continuous rotation in, 55–66, **57–66**
mounting brackets for, **72**, **73**, **73**
overview of project for, 52–66
parts lists for, 68, 95–96, 99
potentiometers in, 54–55
power connector for, 98–99, **100**
printed circuit board fabrication for, 96–98, **97**, **98**
programming and experiments with, 103–116
pulse width setting for, 55, **56**, 67–68, **67**
resistor network in, 62–63, **62**, **63**
servo gear placement in, 65, **65**
servo motor mounts in, 84–89, **85**, **89**
servo motors for, 52, **54**, 54–55
springs and spring-loading mechanisms in, 52, **53**, **54**, 82–83, **83**, 86, **86**, **87**, **88**, **89**
frogbotic.bas/frogbotic.hex, 111–115

gears, gearboxes
Crocobot, 195–198, **196**, **197**, **198**
Frogbotic, 65, **65**
Turtletron, 276, **277**
GOSUB, 33
GOTO, 33

H-Bridge, in L298 dual full-bridge driver, 219–220, **219**, **220**
hacksaw, 1, **2**
hammer, 9, **9**
Harvard architecture, 26
HEX machine code, 39
HIGH, 33
hot glue gun and glue, 7, **9**
HSERIN/HSEROU, 33

I2CREAD/I2CWRITE, 33
IF..THEN..ELSE..ENDIF, 33
In Circuit Debugging (ICD), MicroCode Studio and, 45, 48–49
infrared sensor board, Serpentronic, 148–153, **149**, **151**, **152**, **153**
ircal-serpent.bas/ircal-serpent.hex in, 169–171
adjustment in, 169–170, **170**
calibration of, 154–155, **154**, **155**, 178–179
connector wires and wiring diagram for, 156–157, **157**, **158**

infrared sensor board, Serpentronic *(continued)*
    programming routines for, 177–188
ink-jet printers for PCBs, 17–18
INPUT, 33
input/output (I/O) ports, 25, 27
integrated development environment (IDE) *(See* MicroCode Studio)
ircal-serpent.bas/ircal-serpent.hex in, 169–171

jumping motion, programming and controlling, 110–113, **116**

L298 dual full-bridge driver, Crocobot, 218–221, **219**
laser printers for PCBs, 17–18
LCDIN/LCDOUT, 33
leg assembly, Frogbotic, 77–82, **78–81**
LET, 33
limit switch wiring
    Crocobot, 209, **209**, **210**
    Frogbotic, 91–94, **92**, **93**
limit-switch.bas/limit-switch.hex, 107–110
LOOKDOWN/LOOKDOWN2, 33
LOOKUP/LOOKUP2, 33
LOW, 33

M.G. Chemicals, 18
materials, 12–15
Mecanique *(See* MicroCode Studio)
memory, 25, 26
Metal Supermarket, The, 14
MicroCode Studio IDE, 45–47, **46**
    compiler setup in, 46, **47**
    editor in, 45–46
    In Circuit Debugging (ICD) in, 45, 48–49
    MPLAB programmers and, 46
    one-button compile and programming use, 48, **49**
    PICStart Plus programmers and, 46
    programmer use with, 47–49, **48**
microcontrollers, 25
MicroEngineering Labs, 28
miter box, 1, **2**
mode select switch, Serpentronic, 162, **162**, **162**
motor-test.bas/motor-test.hex, Crocobot, 242–244
motors
    Crocobot, 193
    L298 dual full-bridge driver in, 218–221, **219**
MPLAB programmers, MicroCode Studio and, 46

multimeter, 10–11, **10**, 154–155, **154**

NAP, 33
neural networks, 273–275
NOTEPAD, 35

obstacle avoidance routine, Turtletron, 313–325
ON DEBUG, 34
ON INTERRUPT, 34
one time programmable (OTP) chips, 30
optical shaft encoder (*See also* Turtletron), distance measurement using,
    325–344, **326, 327**
oscilloscopes, 11–12, **11**
OUTPUT, 34

parts lists
    Crocobot, 195, 217–218, 233–234
    Frogbotic, 68, 95–96, 99
    Serpentronic, 120, 145, 150, 155–156, 159
    Turtletron, 276, 284–285, 287, 328
PAUSE, 34
PAUSEUS, 34
PEEK, 34
PIC (*See* programmable interface controller)
PIC 16877, 30
PIC 16C71 in, 232–234, **232**
PIC 16F84, 26–28, **26**, 30
    clock speed of, 26–27
    controller board connection for, **44**
    input/output (I/O) ports in, 27
    pinouts for, 26, **26**
    port A and B connection table for, 28
    registers in, 27
PIC 16F876, 30
PicBasic Pro Compiler, 28–40, **29**
    assembler in, 38
    BAS extension on source files for, 36
    BASIC source files for, 35
    code file in, 38
    command/statements listing for, 32–35
    compiling a program in, 35–40
    EPIC Programmer and, 40–43, **41**
    frog-test.bas program listing for, 36–38
    HEX machine code for, 39
    non-16F84 chips, compiling code for, 38–39

PicBasic Pro Compiler (continued)
  software installation for, 31–35
  testing code in, 38
  text editors and word processors for source file creation in, 35–36
  uncompressing files for, 31
PICStart Plus programmers, MicroCode Studio and, 46
pliers, 6, **6**
PNA4602M module in infrared sensor board, Serpentronic, 148, **151**,
    152
POKE, 34
ports, 25, 30
position sensors, Frogbotic, 91, **91**
POT, 34
potentiometers, Frogbotic, 54–55
power connector, Frogbotic, 98–99, **100**
power supplies, 30, 44
presensitized copper boards for PCB, 19–20, **19**
printed circuit board fabrication, 17–23
  developing, 18–20, **19**
  drilling and parts placement on, 21–22, **22**, **23**
  etching, 20–21
  exposing the board for, 19–20
  ferric chloride for, 18–21, **19**
  Frogbotic, 96–98, **97**, **98**
  presensitized copper boards for, 19–20, **19**
  printers for reproducing, laser and ink-jet, 17–18
  remote control device, 234–239, **235**, **236**, **237**
  resist removal in, 22
  Serpentronic, 146–148, **146**, **147**
  setup for, 19
  soldering, 22
  transparency for, 17–18, **18**
programmable interface controller (PIC), 26
programmer use with MicroCode Studio, 47–49, **48**
pulse width setting, Frogbotic, 55, **56**, 67–68, **67**
PULSIN, 34, 309
PULSOUT, 34
PWM, 34

radio transmitter and receiver modules, Crocobot, 220–221, **221**,
    224–239, **225**, **226**, **238**, **239**
RANDOM, 34
random access memory (RAM), 25, 26
RCTIME, 34
READ, 30, 34

read only memory (ROM), 25
READCODE, 34
README.TXT file, 31
receive-test.bas/receive-test.hex, Crocobot, 252–254
reduced instruction set computer (RISC), 26
registers, PIC 16F84 MCU, 27
remote controller
    Crocobot, 193–194, **194**, 228–239, **229–231**
    Turtletron (*See also* Crocobot), 272–273, **273**, 298–299, **299**
remote-sonar.bas/remote-sonar.hex, 319–325
resist removal from PCB, 22
resistor network, Frogbotic, 62–63, **62, 63**
resistors, 30
RESUME, 34
RETURN, 34
REVERSE, 34
Reynolds Electronics, 220, 299
room mapping, Turtletron, 334–344
rulers and squares, 7, **9**
rx-remote.bas/rx-remote.hex, Crocobot, 257, 258–265

safety glasses, **9**, 10
screwdrivers, 6, **6**
serial data link, Crocobot, 251–269
SERIN/SEROUT, 34, 252
Serpentronic, 117–189
    alternating servo orientation in body sections of, 141, **141**
    assembling mechanical structure for, 137–138, **138**
    battery pack holders for, 128–129, **128, 129, 130,** 139, **139,** 142, **143**
    battery power supply for, 144–145, 158–164
    body section construction in, 121–130, **121–125**
    calibration of infrared sensor board in, 154–155, **154, 155,** 178–179
    connecting body, tail, and head together, 138–143, **140, 141, 142**
    connector wires and wiring diagram for infrared sensors in,
        156–157, **157, 158**
    controller board for, 144–148, **144**
    friction pads for, 163, **164**
    head construction in, 132–136, **133–136**
    infrared sensor adjustment in, 169–170, **170**
    infrared sensor board for, 148–153, **149, 151, 152, 153**
    infrared sensor routine programming in, 177–188
    ircal-serpent.bas/ircal-serpent.hex in, 169–171
    joints in, **120**
    mechanical construction of, 120–121
    microcontroller for, 119

Serpentronic (*continued*)

mode select switch in, 162, **162**, **163**

modifications and customizations for, 188–189, 346

motion programming and control in, 171–176, **172**, **173**, **174**, **175**, **176**, **177**

mounting controller board and infrared sensor board in, 155–158, **156**, **157**

movement of, 119, **120**

movie of, on web site, 189

overview of, 119–136

parts lists for, 120, 145, 150, 155–156, 159

PNA4602M module in infrared sensor board for, 148, **151**, **152**

printed circuit board fabrication for, 146–148, **146**, **147**

programming and experiments with, 164–177

serpentronic.bas/serpentronic.hex for, 179–188

servo horn linkage for, 137, **137**

servo linkage attachment to, 124–130, **125**, **126**, **127**, **128**

servo motor wiring in, 161–164, **161**

servo motors in, 119

size of, 118, **119**

snake biology and, 117–118, **118**

snake-test.bas/snake-test.hex for, 164, 165–169

soldering connections in, 147–148, **148**

tail section construction in, 130–131, **131**, **132**

wiring in, 158–164, **160**, **161**

serpentronic.bas/serpentronic.hex, 179–188

servo motors

continuous rotation modification to, 55–66, **57–66**

control of modified, 66–68

Frogbotic, 52, **54**, 54–55

Frogbotic, control of, 66–68

Frogbotic, modifying for continuous rotation in, 55–66, **57–66**

mounts for, 84–89, **85**, **89**

Serpentronic, 119, 161–164, **161**

SHIFTIN/SHIFTOUT, 35

SLEEP, 35

snake-test.bas/snake-test.hex, 164, 165–169

software installation, 31–35

soldering, 6–7, **7**, **8**, 22

sonar-map.bas/sonar-map.hex, 336–344

sonar-test.bas/sonar-test.hex for, 309–312

SOUND, 35

springs and spring-loading mechanisms, 52, **53**, **54**, 82–83, **83**, 86, **86–89**

SWAP, 35

Tamiya gearboxes
  Crocobot, 195–198, **196, 197, 198**
  Turtletron, 276, **277**
test equipment, 10–12
testing controller boards, 44–45
text editors and word processors for source file creation in, 35–36
Thinkbotics web site, 17
timers, 25, 30
TOGGLE, 35
tools, 1–10
transmit-test.bas/transmit-test.hex, Crocobot, 254–256
TRIS register, PIC 16F84 MCU and, 27
turtle-receive.bas/turtle-receive.hex, 300, 301–305
turtle-trans.bas/turtle-trans.hex, 300, 305–308
Turtletron, 271–344
  avoidance.bas/avoidance.hex for, 313–319
  base for (Frisbee), 279–281, **279, 280, 281**
  battery pack holder for, 296–297
  controller board and electronics of, 283–286, **284**
  cover supports for, 282, **282**
  differential drive system in, 283
  distance measurement using optical shaft encoder in, 325–344, **326, 327**
  encode-test.bas/encode-test.hex for, 332–334
  gearbox assembly and attaching wheels to, 276, **277**, 277–283, **278**
  history of robotic turtles and, 273–275, **274**
  mechanical construction of, 275–299
  modifications and customizations for, 347
  obstacle avoidance routine for, 313–325
  optical shaft encoder in
    encode-test.bas/encode-test.hex for, 332–334
    fabrication of, 327–334, **328, 329, 330, 331**
    mounting of, **331**
    parts list for, 328
    room mapping using, 334
    sonar-map.bas/sonar-map.hex for, 336–344
    wiring diagram for, **332**
  overview of, 272–273
  parts lists for, 276, 284–285, 287
  programming, 300–325
  remote controller for (*See also* Crocobot), 272–273, **273**, 298–299, **299**
  remote-sonar.bas/remote-sonar.hex for, 319–325
  room mapping for, 334–344
  sonar-map.bas/sonar-map.hex for, 336–344

Turtletron (*continued*)
    sonar-test.bas/sonar-test.hex for, 309–312
    turtle and tortoise biology and, 271–272, **272**
    turtle-receive.bas/turtle-receive.hex for, 300, 301–305
    turtle-trans.bas/turtle-trans.hex for, 300, 305–308
    ultrasonic range finder in, 272, 286–298, **286**
    antenna attachment to, 295–296, **296**
    avoidance.bas/avoidance.hex for, 313–319
    connecting to robot, 290–291, **290**, **291**
    connections for, 288, **288**
    housing for, 292, 293, **293**, **294**
    neck mount for, **294**, **295**
    obstacle avoidance routine for, 313–325
    operation of, 288
    pulse width setting in, 309
    remote-sonar.bas/remote-sonar.hex for, 319–325
    room mapping using, 334–344
    sonar-map.bas/sonar-map.hex for, 336–344
    sonar-test.bas/sonar-test.hex for, 309–312
    testing, 308–312
    timing of, 289, **289**
    wheels for, 277–283, **278**
    wiring diagram for, 297
tx-remote.bas/tx-remote.hex, Crocobot, 257, 265–269

Ultraedit, 35
ultrasonic range finder (*See* Turtletron)
ultraviolet erasable chips, 30
uncompressing PicBasic Pro files, 31

vise, 1, 3, **3**
Von Neumann architecture, 26

walk-routines.bas/walk-routines.hex, Crocobot, 245–251
Walter, William Grey, 273–275
WHILE..WEND, 35
wire strippers, 6, 7
wiring
    Crocobot, 201, **201**
    Crocobot, transmitter, **238**
    optical shaft encoder (Turtletron), 332
    Serpentronic, 158–164, **160**, **161**
    Turtletron, 297
word processors for source file creation in, 35–36
wrenches, 6, **6**

WRITE, 30, 35
WRITECODE, 35

XIN, 35
XOUT, 35

About the Author

Karl Williams is an independent robotics researcher, electronics experimenter, and software developer. He is with Mitra Imaging, a leading medical imaging software company just acquired by AGFA HealthCare Informatics. He is the author of *Insectronics—Build Your Own Walking Robot* and has written for *Nuts and Volts* magazine. A resident of Ontario, Canada, Karl has hosted a robotics and electronics Web site for four years, and received an IBM regional computer technology award for building a computer-controlled robotic arm.